AGRICULTURAL EXTENSION

AGRICULTURAL EXTENSION

Nekesah T. Wafullah

Delve
PUBLISHING

www.delvepublishing.com

Agricultural Extension

Nekesah T. Wafullah

Delve Publishing
4164 Lakeshore Road
Burlington, ON L7L 1A4
Canada
www.delvepublishing.com
Email: orders@arclereducation.com

ISBN: 978-1-77956-402-3 (Hardcover)

ABOUT THE AUTHOR

Nekesah T. Wafullah is a skilled agriculture expert with extensive knowledge in agricultural energy value addition products, agricultural business management services, project management, various forms of fertilizer, their production, sales, marketing aspects and application regimes; cross border fertilizer trade policies; youth and women empowerment and volunteerism. She is adept at project planning and management as well as creating simple solutions to complex problems. She has experience within agricultural markets in Kenya, Zambia, Malawi, Tanzania, Rwanda, the Democratic Republic of Congo (DRC)- Bukavu and Lubumbashi and Uganda. She mentors high school and college students and advocates for better performance in Agricultural science. During her free time, she loves editing books, watching movies, cooking, baking, networking, reading and dancing. Nekesah holds an M Sc. In Agricultural and Applied Economics degree from the University of Nairobi with a major in International Trade and Policy.

TABLE OF CONTENTS

LIST OF FIGURES

LIST OF TABLES

LIST OF ABBREVIATIONS

AAS	Agricultural Advisory Service
ADAS	Agricultural Development Advisory Service
AFADA	Albanian Fertilizer and Agri-Business Dealers' Association
AIDS	Immunodeficiency Syndrome
AKIS	Agricultural Knowledge and Information System
AKIS/RD	Agricultural Knowledge and Information Systems for Rural Development
CES	Community Extension Services'
CREA	Council for Agricultural Research and Economics
EU	European Union
FAO	Food and Agriculture Organization
FCE	Fee-Charging Extension
FFMS	Farmers' Forest Management Schools
FFS	Farmer Field Schools
FSD	Farming Systems Development
FSR	Farming Systems Research
GMOs	Genetically Modified Organisms
HEP	Health Extension Program
HEWs	Health Extension Workers
HIV	Human Immunodeficiency Virus
HRD	Human Resource Development
ICT	Information Communication Technology
IPM	Integrated Pest Management
IT	Information Technologies
KAP	Knowledge, Attitude, and Practice
M&E	Monitoring and Evaluation
MOA	Ministry of Agriculture
NAADS	National Agricultural Advisory Service

NAESRI	National Agricultural Extension Systems Reform Initiative
NEPAD	New Partnership for Africa's Development
NGOs	Non-Governmental Organization
OECD	Organization for Economic Co-operation and Development
PRA	Participatory Rural Appraisal
SDRE	Extension, Education, and Communication Service
SEAGA	Socio-economic and Gender Analysis
SPFS	Special Program for Food Security
T&V	Training and Visit
TI	Technical Innovation
WHO	World Health Organization
WTO	World Trade Organization

ACKNOWLEDGMENT

The book is the product of great effort and time. Its completion would not have been possible without the participation and assistance of so many people, whose names may not all be enumerated. Their contributions are sincerely appreciated and gratefully acknowledged. However, I would like to express my deep appreciation and indebtedness, particularly to the following: Arcler Education, Inc., and Charles M. Kuria for their endless support, kind and understanding spirit during this undertaking. To all relatives, friends, and others who, in one way or another, shared their support—whether morally, financially, or physically—I would like to personally thank you.

PREFACE

Extension is an important tool in development. This is because extension is a function that can be applied to various areas of society. It operates in the industrial, health, and education sectors, as well as agricultural and rural development. Originally derived from university extension, the term extension is therefore applicable to various areas of development. Extension programs are normally believed to be the services that are meant to extend research-based knowledge in various areas, such as agriculture, industry, education, and so forth, to improve the living conditions of the individuals. To achieve the best outcome from utilizing extension, certain principles need to be put into consideration. One such principle is that extension work best with rural people in need of household improvement. Therefore, to help with the required improvement the extension agent needs to consider extension as a two-way link that requires cooperation from all stakeholders.

In any society there are strong pressures on its members to behave in certain ways. For the farmers, some of these pressures will come from within. In all societies there are accepted ways of doing things and these ways are directly related to the culture of the society. These societal pressures greatly affect extension efforts, thus requiring the agent to take into consideration factors like age, education, marital status and income levels when spreading extension programs. Another factor that affects the implementation of the extension services is the agents' skills such as communications, preparedness, composure, and education.

Several approaches are applied in extension programs. There is no one universal type of extension but a variety of activities and approaches which can be called extension. The appropriate extension methods have been classified into three categories: mass, group, and individual extension methods. Each method has its advantages and disadvantages. While group and mass methods reach a wide audience and are relatively cheap, they are not as personalized as individual extensional methods, which on the other hand are relatively expensive. Despite the advantages and disadvantages none of the methods is superior to the other, the use is dependent on the context.

The various kinds of challenges that have been encountered in extension services are related to the usage of technologies, marketing, inputs, resources, finance, education, training, and awareness, managerial skills, health conditions and natural resources and environmental conditions. These challenges are mitigated by extension program reforms. Reforms are done after a program is evaluated and recommendations are given for change depending on what worked and what needs changing. Several guidelines are followed in carrying out an extension reform for best case outcome. This book gives case studies of extension programs, there potential, lessons, and impacts.

Chapter 1

EXTENSION

Contents

1.1. INTRODUCTION

Broadly speaking, extension is a function that can be used in many societal contexts. It works in the fields of agriculture and rural development, as well as industry, health, and education. Since the term "extension" was first used to refer to university extensions, it can be applied to many different types of development. Israel (1982) emphasized in previous conversations on the Training and Visit (T&V) extension system supported by the World Bank that the principles of T&V might be extended to other sectoral systems that are involved in providing nonformal education. But it also cautioned that T&V was built on traditional management ideas that would probably not work in poor nations. This lesson wasn't completely understood until the 1990s, and it's partly responsible for the present focus on participatory management ideas. Extension programs are seen to be services designed to disseminate research-based knowledge in a variety of fields, including education, industry, and agriculture, with the goal of enhancing people's quality of life (Table 1.1). Technology components, transformation, aims for complete rural development, managerial abilities, and non-formal education are all included in extension. The conventional understanding of extension in developing nations was primarily concerned with raising production, enhancing yields, educating farmers, and dispersing technologies. In the modern era, extension is understood to go beyond the transfer of technology for facilitation and training to learners. It also involves addressing market concerns and collaborating with a wide range of service providers and other organizations. Obtaining the knowledge, abilities, and other technologies necessary to better their standard of living is the focus of extension services provided by organizations (Davis, 2009).

Table 1.1. *Extension as a Function in Various Sectors of Society*

Education	**Agriculture**	**Rural Development**	**Health**	**Industry**
University Extension (Continuing Education)	Agricultural Extension	Rural Development Extension	Health Extension Services	Industrial Extension

Source: William et al., (2001).

Traditionally, extension has been crucial in disseminating knowledge and encouraging new technologies or approaches to production and manufacturing management. The main component of extension services is the application of cutting-edge, contemporary strategies and procedures. Additionally, extension serves as a link between researchers and other innovation system participants and farmers, manufacturers, and other workers.

1. **Extension Services in Education:** Extension is a crucial part of a university that uses teaching and research to inform its goals and initiatives for providing community services. Building relationships via education is the foundation of community extension work. The goal of Community Extension Services' (CES) initiatives and programs is to assist individuals in empowering themselves through long-term initiatives. Through services that transfer knowledge and technologies to create an influence on holistic and sustainable development programs and people empowerment, extension enables educational programs to reach and "touch the lives" of communities and the environment. The most valuable project at CES is called "HELPS," which stands for human needs. These needs include physiological ones, which can be met by maintaining good health; safety and security needs, which can be met by a strong environmental program; and literacy needs, which can be met by improving one's self-worth and confidence and helping one become a productive person. If these comprehensive facets of human nature are given high-quality, long-, and short-term programs and services, they will eventually help create a family that is sustainable, which will then lead to a community. Through the provision of services consistent with the organizational plan and activities developed by each department in line with the institution's vision and mission, this initiative aims to meet the needs of the adopted community. It relates specifically to each department's staff and students' areas of competence.
2. **Extension Services in Health:** Health extension projects are neighborhood-based endeavors with the goal of enhancing healthcare outcomes and access, frequently by assigning skilled personnel to supply necessary services. Health Extension Workers (HEWs) are employed by the Ethiopian Health Extension Program (HEP) to provide primary and preventive healthcare services, with a special emphasis on mother and child health in rural regions. By implementing various health extension packages, these initiatives prioritize preventative, promotive, and basic curative interventions to improve the health status of families. Furthermore, Health Extension Regional Officers (HEROs) in the US serve as a liaison between community health needs and research agendas, promoting community involvement and research congruence to tackle regional health issues. The idea of health extension has also been applied to primary care settings, where outside assistance is given to support efforts to improve quality and spread innovation. This illustrates the

adaptability and efficiency of health extension models in a variety of contexts. Healthcare workers are expected to acquire high levels of competency in order to handle complicated health issues and meet the diverse requirements of people in the current environment of change and uncertainty. Furthermore, the assumption of safety, quality, and accessibility improvements in healthcare is the foundation for the competences of healthcare workers, including leadership, clinical reasoning, ethical qualities, and successful cooperation.

3. **Extension Services in Industry:** Extension services are a typical method for offering small-scale businesses a variety of assistance. These support business owners in recognizing opportunities, recognizing issues, evaluating their own strengths and limitations, and offering prompt guidance on all issues. There are significant differences in the focus of extension services, how they are run, and how much they are linked to further financial or technical support.

   The term "extension" can be interpreted in a lot of different ways. Most likely, every extension agent has a unique definition of what an extension is. This comprehension will be predicated on prior knowledge as well as the specific kind of extension service the agent is working with. Put otherwise, there isn't a single, widely accepted definition of extension that can be used in every circumstance. Moreover, the concept of extension is dynamic as its meaning is subject to constant modification. Therefore, it is not possible to define extension precisely; rather, it refers to a continuous and dynamic process that occurs in rural areas. Examining several remarks that have been written about the term extension will help.

- An informal educational approach aimed at the rural populace is called extension. This procedure provides guidance and knowledge to assist people in resolving their issues. Extension seeks to raise a farm family's standard of life overall as well as the productivity and efficiency of the family farm.
- The aim of extension is to alter farmers' perspectives regarding their challenges. Extension addresses not only the material and financial gains but also the personal growth of the rural populace. Therefore, extension agents engage in conversation with rural residents to assist them better understand their issues and determine how to address them.
- The process of working with rural residents to enhance their standard of living is known as extension. This entails assisting

farmers in raising the productivity of their farming while also strengthening their capacity to guide their own personal growth.

The claims are provided to show the variety of views on extension that exist. Nonetheless, there are certain things in common amongst them. They all emphasize that extension is not a singular, one-time event, but rather a process that takes place over time. Additionally, they all emphasize that extension is a kind of education that works with and supports rural communities, empowering them to face challenges head-on. Four key components can be found in the extension process if claims like the ones above are looked at more closely and the concepts and practices of the present day are considered:

i. **Knowledge and skills:** Extension can provide farmers with additional knowledge and information that they do not already possess, even when they are highly knowledgeable about their farming systems and environments. The agent can help farmers by imparting knowledge on topics such as broad concepts of pest control, the reasons behind crop damage, and the processes involved in the breakdown of compost and manure to supply nutrients to plants. By applying this knowledge, farmers often need to acquire new skills of various kinds, such as technical skills to operate unfamiliar machinery, organizational skills to oversee a group project, the ability to weigh the financial benefits of technical advice, or farm management skills to maintain records and allocate the use of equipment and resources. An essential extension activity is transferring knowledge and skills to farmers and their families, for which the extension agent must adequately prepare. The agent must identify knowledge or skill gaps among the farmers in the region and then set up appropriate educational opportunities to help them address these gaps.

ii. **Technical Advice and Information:** Additionally, extension offers guidance and data to help farmers make decisions and, in general, empower them to act. This can include details about marketplaces and prices as well as the accessibility of financing and inputs. The technical guidance will most likely have a greater direct impact on the family farm's production processes and the actions required to enhance or maintain this output. The results of agricultural research will serve as a major foundation for much of this technical guidance. But in many cases, farmers may also provide other farmers with insightful counsel and information, so agents should constantly try to build a connection amongst farms.

iii. **Farmers' Organization:** Farmers require some kind of organization in addition to knowledge, data, and technical guidance in order to represent their interests and provide a platform for taking coordinated action. Therefore, the goal of extension should be to assist in the formation, growth, and development of local farmer organizations. This ought to be a cooperative endeavor, and any such establishment should come about only after the farmers have been consulted. These groups will facilitate communication between extension services and nearby farmers in the future and act as a conduit for knowledge and information sharing.

iv. **Motivation and Self-Confidence:** Isolation and the belief that there is little that farmers can do to improve their lot in life are two major obstacles to development that many of them encounter. Some farmers will have struggled their entire lives under trying conditions to support their families with little help or encouragement. It is critical that extension collaborates closely with farmers, supporting their initiative and generally enticing them to participate in extension operations. Convincing farmers that they are self-sufficient, capable of making their own decisions, and capable of escaping poverty is equally crucial.

The four main components of the extension procedure are listed above. It is not implied that any one of these components is more significant than the others or that all extension initiatives must include them. Obviously, the specific conditions will dictate the extension strategy. Nonetheless, these components should form the foundation of a comprehensive extension agency, which should work to promote them in rural regions. There are times when the issues facing the local farmers will call for quick information and guidance; other times, more careful planning and inspiration may be needed. These many requests must be met by an extension service.

1.2. PRINCIPLES OF EXTENSION

To carry out extension projects and programs, the majority of governments in the developing countries have established officially structured extension services.

The money, offices, staff, and other resources all support the extension activity. Prior to delving deeper into extension in subsequent chapters, it will be beneficial to think about the guiding concepts.

1.2.1. Extension Works with People, Not for Them

Extension serves rural residents. An extension agent does not try to make decisions for people; only the individuals themselves can decide how they will farm or live. Given complete information, including potential alternatives to solve their problems, rural residents can make informed judgements about their situation. Making choices helps people become more confident. Therefore, extension disseminates information, aids in problem solving, and motivates farmers to make choices. Individuals are more confident in judgements and programs they have created for themselves than in ones that are forced upon them.

1.2.2. Extension Is Accountable to Its Clients

There are two sets of master's for agents and extension services. They answer to the government ministries that set policy for rural development as well as their senior officers, on the one hand. Agents must conduct their business in accordance with government regulations and procedures. Extension, on the other hand, serves rural communities as their servant and is tasked with meeting local requirements. This implies that the rural poor ought to be involved in determining the true effectiveness of extension. Examining the degree to which plans and policies have been implemented is one way to gauge their efficacy. The degree to which extension work has raised rural residents' earnings and standards of living is a similarly significant indicator. Therefore, in addition to technical and national economic objectives, extension activities are focused on the needs of the public. It is the extension agent's responsibility to coordinate these demands. For instance, increasing the quantity of food produced and sold in the nation may be a key component of government policy. The needs of the country and the farmers themselves can be met by opting to support most small farmers in raising their output through better farming practices.

1.2.3. Extension is a Two-Way Link

The process of extension is two-way; farmers and their families receive knowledge and ideas from the extension agent. These recommendations, which are frequently based on research from agricultural and other stations, are undoubtedly valuable, but so is the information that farmers provide to extension and research personnel. Extension should be prepared to both give and accept ideas, comments, or advice from farmers. This reciprocal exchange of ideas might take place at many phases. during the problem-defining process. Maintaining a consistent communication line with farmers allows the extension agent to assist researchers in comprehending local farming issues and constraints. To make sure that research recommendations are pertinent to farmers' requirements, it is even better if the agent can put researchers in direct contact with farmers. Farmers and research are connected through extension. when suggestions are put to the test in practical settings. While a novel agricultural technique or crop variety may yield positive outcomes in a research facility, it may not perform as well in a farmer's field. Research personnel can receive input and test recommendations on farmers' fields through trials. the moment farmers implement advice. Farmers occasionally find issues with a recommendation that the research station overlooked. The recommendations can be modified in light of the input. A essential component of good extension practice and one that ought to guide all extension efforts is the two-way relationship that exists between the farmer, extension, and research.

1.2.4. Extension Cooperates with Other Rural Development Organizations

Extension agents and services should collaborate closely with other organizations in rural regions that offer vital services to farmers and their families. Extension is but one facet of the numerous political, social, and economic initiatives aimed at improving rural society. Therefore, when preparing to implement extension policies, Extension must be ready to work with all other similar organizations, both government and non-government. The following are the kinds of organizations that extension services ought to collaborate with:

1. Political organizations and local politicians who actively support the extension agent locally will be helpful in putting the agent in contact with local farmers.
2. Support groups include those that offer marketing services, finance facilities, or inputs for agriculture or other industries. If these inputs are to be useful at all, they need to be accessible in adequate amounts, at the appropriate locations, and at the appropriate times.
3. Health services, to keep the extension agent informed about health issues in the area, especially those related to nutrition. Nutrition and agricultural development are strongly intertwined; thus the agent must stay up to date on health-related initiatives and programs and modify his plan to meet regional health standards.
4. Local schools, to give the agent early access to the next generation of farmers and start preparing them with the knowledge and abilities needed for farming.
5. Community development, whose goals will resemble extension's educational efforts in a big way. To support community action

initiatives and remove social and cultural barriers to change, extension agents frequently collaborate closely with community development professionals in the area.

6. It is crucial that the field extension agent understands what his counterparts in other government agencies and services are doing, as well as what they are doing. In addition to preventing redundancy, close collaboration opens up possibilities for integrated farm initiatives.

1.2.5. Extension Works with Different Target Groups

Extension is aware that not every farmer in a given region would have the same issues. Some will be eager to test out new ideas and will have access to more land than others. Those who possess fewer resources are likely to exercise greater caution. Extension is unable to provide a single "package" of guidance that works for every farmer. The agent must identify several groups and create programs that are suitable for each group.

A large portion of extension work in the past was focused on the progressive farmer who was supposed to introduce new concepts to other farmers. However, it has been seen that this is not always effective because progressive farmers frequently face distinct issues.

They are typically more involved in the marketing of their produce, possess more land, and have higher levels of education. As a result, Extension needs to consider the existence of various farming groups while designing its programs. Since they might not have the means to participate in extension efforts, the smallest and poorest farmers will require extra care. It is important to emphasize that diverse farmer groups exist in every community, each with unique resources and abilities, and that extension must adapt to these groups appropriately.

1.3. EXTENSION AND EDUCATION

It has been observed that the role of the extension agent involves education. To enhance farming and other productive endeavors, farmers, and their families must acquire new abilities, know-how, and methods.

As a result, they acquire new perspectives on farming, the new methods, and extension itself, which in turn shapes their conduct going forward. However, extension agents also need to be ready to learn about farming practices from farmers and stay current with pertinent advancements in agricultural knowledge. The agent in this extensional educational work should understand several learning concepts.

1.3.1. The Educator Must Also Be a Learner

The goal of education is not to stuff knowledge into vacant minds. Farmers would not be able to survive if they did not already possess a wealth of knowledge about their farming system and surrounding environment. Extension needs to expand upon existing knowledge. Therefore, in addition to teaching, an extension agent must also learn. He needs to study what farmers already know about agriculture, such as how they explain and describe events that occur on their properties and what they already know about better farming practices.

1.3.2. Learning Requires Motivation

Nobody has the right to force someone else to learn. A desire to study must exist. It is harder for adults to understand new concepts and information than it is for kids. Additionally, the extension agent does not have a captive audience like a teacher does. Farmers have the freedom to ignore extension agents' advice and to not learn. Individuals don't learn unless they believe that what they are learning will enable them to fulfil a need or want. In addition to being necessities for survival, food, and drink are also important for establishing and growing a family and looking for secure housing that shields the family from harm and discomfort. People have needs, but they also have wants or desires. Though not as urgent as demands, these are nonetheless significant. People want respect and admiration from their friends and family, as well as status in their community and the goodwill of their neighbors. After basic requirements are met, these desires become more obvious. Despite having decent homes and access to food, farmers, and their families are constantly looking to improve. Their goal is to increase production, and the extension agent can help them achieve this goal by assisting them in improving agricultural practices. This is a valid goal that will increase farm productivity. A driven farmer will probably pick up new skills faster and more thoroughly than an unmotivated farmer. It is imperative that the extension agent keeps in mind this crucial aspect.

1.3.2.1. Dialogue and Practice Are Important for Learning

A group of farmers is instructed by an extension agent on how to thin their crops in order to increase yields. Then, believing the farmers had mastered the new ability, he leaves. When he returns a few weeks later, none of the farmers have thinned their crops, and what he told them is only vaguely remembered. It is not surprising that the extension agent is here. A plain conversation does not teach farmers much, and the majority of what they do learn they quickly forget. However, a lot more will be learnt and retained if they are given the opportunity to ask questions, explain the new material in their own terms, and have a conversation with the extension agent. Moreover, farmers need to be given the opportunity to practice newly acquired practical skills. The farmer will next acquire the confidence to apply the new skill once the extension agent can fix any initial errors.

1.3.2.2. Learning and Adoption Occur in Stages

Extension incorporates several learning modalities. A group of farmers must become aware of a new practice before they can decide to give it a trial. Then, they might need to pick up some new abilities. The process of absorbing novel concepts can be divided into five phases.

1. **Awareness:** A farmer finds out about the concept's existence but lacks knowledge about it.
2. **Interest:** The farmer becomes intrigued by the concept and asks an extension agent or a friend for further information.
3. **Evaluation:** It is now necessary to take the farmer's impact into account. What benefits will it provide? What drawbacks or challenges does this novel concept present? The farmer can decide whether to try out the new idea after gathering more information or by attending a conference or demonstration.
4. **Trial:** Farmers frequently choose to give the concept a trial on a smaller scale. For instance, they might choose to treat a tiny portion of one field with fertilizer or manure and then compare the outcome to the remainder of the field. To do this, they ask for guidance on the best times and methods for applying manure or fertilizer.
5. **Adoption:** If the farmers are persuaded by the experiment, they fully embrace the concept and incorporate it into their traditional farming practices.

Individual farmers, farmer organizations, or entire communities go through similar phases. Within communities and groups, the process is more intricate and might take a lot longer. The extension agent provides the appropriate information and assistance at each stage of the process by utilizing a variety of extension techniques. He has to set up educational opportunities that advance them to the next level. He starts out a community forestry program by urging people to acknowledge that there is an issue with the number of trees declining and that this may be solved by the community planting and maintaining a wood lot. A visit to another village that has previously established a wood lot can then heighten interest. There will be a lot of discussion in the village throughout the appraisal phase. The extension agent is available to address concerns and offer comprehensive details regarding the expenses and benefits. He can set up skill-training sessions when a decision is made to act.

1.3.2.3. Farmers Differ in Their Speed of Learning and Adoption

Diffusion is the process by which a novel concept gets adopted by locals. Not every farmer will embrace a novel concept at the same moment. In every rural community, farmers differ in their willingness to adopt and implement new ideas based on a variety of factors, including their personality, the quantity of land and other resources available, and their prior experience implementing new ideas.

As a result, we may classify farmers according to how well they are able to implement new concepts. pioneers. Farmers that are open to new ideas are innovators. This class often consists of a small number of individuals in a rural community. They are typically farmers who feel free to make their own decisions without concern for other people's opinions because they have lived outside of the village for a while. Innovators are frequently met with mistrust and jealousy in rural communities.

Nonetheless, they are critical to an extension program's success because they might be convinced to test novel approaches, raising public awareness of them in the process. The extension agent should, however, use discretion and tact and refrain from publicly praising innovators excessively or spending an excessive amount of time with them. The rest of the community may then reject the concept out of envy and mistrust of the innovator's motivations for trying new things.

1. **Early Adopters:** Early adopters are farmers who are more circumspect and prefer to see the concept tested and validated in their own environment. Although they show early curiosity, they must first be persuaded of the idea's direct benefits through the presentation of results. This group of farmers typically consists of respected members of the community as well as local leaders.
2. **The Majority:** The remainder of the farmers will likely adopt a new idea more slowly and maybe less fully if it is adopted. Many farmers won't have the resources to implement the new concept as all, and those who do might only do so cautiously and slowly. The opinions of neighbors and local leaders are likely to have a greater impact on most people who can and will adopt the notion that will the extension agent or the demonstrations he organizes.

1.4. TYPES OF EXTENSION

There are many different actions and methods that fall under the umbrella of "extension," rather than a single, all-encompassing kind. Agricultural extension is the most prevalent kind of extension in rural regions, as it has already been mentioned that agriculture is the foundation of a rural economy. However, farming is not the only field in which farmers and their families need fresh ideas and areas of knowledge.

There exist additional facets of family life that can be enhanced by novel insights and approaches. Any program that works with farmers and their families to enhance their social and economic circumstances and strengthen their capacity to take charge of their own future development is known as extension. But there are a few various ways this extension might happen, so it's good to go over the two main ones first.

1. **Agricultural Extension:** More extension agents are more likely engaged in agricultural operations than in any other area of rural life. This emphasis on agricultural extension makes sense given the significance of agriculture and the requirement to generate food for the farm family as well as the country at large. While some agricultural extension programs focus primarily on a single crop, others take a broader "whole farm" approach. The decision is heavily influenced by the national crop requirements as well as the local agricultural system. Single crop expansion is more typical in areas where cash crops like cotton, cocoa, or sugar are grown. In addition to providing farmers with the services and inputs they need to support their agricultural output; an agricultural extension service provides expert advice on agriculture. It informs farmers and disseminates fresh concepts created by agricultural research stations. Programs for agricultural extension encompass a wide range of topics, such as enhanced crop varieties, more effective management of livestock, better water management, and the prevention of pests, weeds, and plant diseases. In certain cases, local farmers' associations and organizations may be strengthened through agricultural extension, enabling them to get access to extension services. Therefore, agricultural extension offers farmers the essential components they require to raise their level of agricultural productivity.
2. **Non-Agricultural Extension:** It is convenient to refer to all other forms of extension as non-agricultural extension since there isn't a single term that can be used to describe them all. This phrase refers to all endeavors and activities that are significant to farm families but are not specifically connected to agriculture or the raising of livestock. Non-agricultural extension efforts include population education, community development, home economics, and family health and nutrition.

1.5. SOCIAL AND CULTURAL FACTORS IN EXTENSION

Farmers are part of the society in which they live, as do their families. Every civilization places a lot of pressure on its people to act in particular ways. Some of these pressures will come from within the farmers. Every society has norms for behavior, and these norms are closely tied to society's cultural norms. The cultural norms of their society have an impact on the attitudes and wants of farmers. People will grow up believing that a farmer's method of scattering seed and ploughing it into the soil is the sole proper way to plant if that is the custom in that community. Despite being informed of the advantages of alternative approaches, their deeply ingrained beliefs could be resistant to change. However, not all of these pressures will originate from the attitudes and beliefs of the farmers; some may come from other sources. Every society has expectations about how its members should act. Nobody is perceived by others as being alone. Every individual is perceived as holding a certain position in society, and these positions come with expectations. Unmarried men are expected to labor on their father's farm in some communities; they are not expected to start farming their own plot until they get married. A prosperous farmer could be expected to support less fortunate relatives with food, cash, and housing, or to cover the cost of his relatives' children's education. A person will face criticism from others around him if he defies these standards. Most people behave in ways that are consistent with these expectations because they enjoy feeling accepted and approved by others around them. Understanding the social and cultural background of the farmers he deals with will make an extension worker more effective. Subsequently, he will be more adept at providing counsel that meshes with the society's culture and can leverage the organization and customs of the society to advance his professional goals. Therefore, it is helpful to look at the key aspects of cultures and societies that are pertinent to extension work.

1.5.1. Social Structure

A society's structure refers to how its members are arranged into families, tribes, communities, and other divisions or groups. The groups to which an individual belongs affect not just that person's access to opportunities, jobs, and land, but also that person's attitudes and the expectations of others towards them.

1.5.1.1. Social Divisions

A society may be divided due to a variety of reasons, such as age, sex, religion, place of residence, kinship, and shared economic interests.

1. **Age:** Individuals that are of the same age typically share similar hobbies and viewpoints. Compared to older people, young people typically have different beliefs, attitudes, and life goals. Elderly people are respected, and their opinion is highly considered in many societies. An extension agent must become familiar with the specific goals, standards, and limitations of the various age groups within the community in which he operates.
2. **Sex:** Traditionally, men, and women in rural areas have performed different tasks. Typically, women do domestic tasks like cooking, gathering firewood and water, and taking care of the kids. Nonetheless, women are also heavily involved in farming in many nations. Women typically perform more than 60% of the agricultural labor in several African nations. Typically, men tend to commercial income crops like oil palm or tobacco, while women tend to home farms where they raise food crops. In other places, men, and women perform separate tasks in the same fields. For instance, in Botswana, women perform weeding, bird-scaring, and threshing, while men are historically responsible for ploughing and all cattle-related tasks. Male extension agents visit male farmers, focusing agricultural extension efforts mostly on men. However, unless extension agents include women in their programs, any changes made to the farming practices will also have an impact on women, and they may not succeed.
3. **Religion:** Religious group members often share attitudes and beliefs, which may have an impact on how open they are to collaborating closely with members of other religions. The extension agent should be mindful of the potential for conflict in rural communities due to religious differences. Certain faiths enforce behavioral norms that may have an impact on extension. Farmers may be unavailable for farm labor or extension operations during specific times of the day, specific days of the week, or specific seasons of the year due to religious festivals.
4. **Residence:** Nearby residents typically have certain interests with one another. A village's citizens will desire amenities like a school, potable water, and medical care. They'll demand equitable participation in government development initiatives as well as road access. The village can come together around these shared interests, especially if those interests are in danger. Extension agents must to make every effort to incorporate into their programs activities that would bring the community together around a single goal. However,

they ought to be conscious that a village might also have factions. For instance, some villagers may argue that a new water tap should be placed close to their homes, while others may feel that it should be placed elsewhere in the hamlet. Extension agents should, to the greatest extent feasible, refrain from exacerbating any existing tension within a community and instead look for ways to ease it. It may be exceedingly difficult for an agent to be effective if it is perceived that he is representing a specific group in the village.

5. **Kinship:** Within and within families, the strongest groupings are frequently those based on birth and marriage links. A man, a woman, and their offspring make up the family, the smallest of these units. These are autonomous families that choose where to live, where to farm, and what kinds of crops to plant in certain civilizations. However, these families will typically be expected to satisfy specific obligations to close relatives, which may limit their freedom of choice. Larger familial groups may cohabit, share land, or even decide on farming jointly in other communities. The individual farmer may not have much decision-making freedom in this situation. It would be necessary for an extension agent to establish close working relationships with the leaders and decision-makers of these groups.

6. **Common economic interest:** Economic disparities play a significant role in social structure. A person's occupation, income, and the kind of land they own or may rent are all variables that can create social divisions within society, each with its own issues, passions, values, and attitudes. Landowners and renters, commercial farmers interested in learning about the newest farm equipment, subsistence farmers unable to purchase expensive inputs, and livestock and grain farmers can all be found in a rural region. Every group will have distinct needs and demands from the extension agent, and the agent must tailor his approach to suit the goals of each group. The quantity of land and money that each farmer possesses are the primary economic elements dividing rural societies. This is especially evident in some Latin American nations, where the majority of families either work as agricultural laborers or cultivate their own tiny plots for subsistence, while a tiny percentage own extremely vast estate. Together with small and large farmers, most Asian countries also have a sizable population of landless laborers. Regarding extension support, these farmer types differ greatly from one another. The kind of guidance and assistance that an extension agent should provide to

each farmer category varies depending on their economic situation. Such distinctions impact a farmer's relative economic and political power as well as the standard of living that individuals can attain. Because they buy and sell in greater numbers, large farmers are more likely than small farmers to be provided credit, and merchants and dealers will offer them better conditions. They are frequently more easily listened to by planners and political leaders. Working with the larger farmers may also appeal to extension agents more. Extension agents must recognize these differences and find strategies to assist farmers who are eager to make improvements to their farms but lack significant political or economic clout if they are to assist smaller farmers.

1.5.1.2. Groups

It has been explored how a society's large social divisions impact the attitudes, needs, and interests of its people. Every society also has small groups of individuals that meet for a shared activity or goal. A few of these organizations might endure for a very long period. For example, a savings club might meet every week for years on end. Club membership may fluctuate, but it will always exist. Other organizations might only last a short while, such when a few neighbors decide to pitch in to help each other with farm chores on their property. These groups frequently serve as the foundation for extension groups and can be extremely helpful for extension agents. For instance, in the Republic of Korea, Mothers' Clubs—which have a significant impact on village development initiatives—have evolved from traditional women's savings groups. These groups provide labor for self-help initiatives, earn substantial funds for community projects, and serve as a resource for rural women seeking popular education and farming knowledge. However, people's willingness to join groups varies. Kinship groups may collectively hold land in certain areas, but each tiny family is left to farm its own plot. Any modification to farming practices in this scenario would require independent judgements from numerous farmers. In some places, the kinship group members might work on the land jointly. In that scenario, an extension agent ought to collaborate with the group to enhance farming practices, most likely through its traditional leaders. When feasible, an extension agent should utilize these groups, and in order to help him with his task, he should try to find out:

- What groups there are in the region,
- What joint activities are undertaken by members of each group,
- The interests of each group, and

- Whether any of the groups could form the basis of an extension group.

1.5.1.3. Formal and Informal Leaders

Men and women who make decisions on behalf of others or who are revered by others exist in every society, and as such, they have some influence over the attitudes and actions of others. These leaders may play a critical role in the accomplishment of extension initiatives. Formal leaders are those who occupy positions of acknowledged authority. Once one understands the pattern of leadership in the culture, they are typically easy to spot. A person in a position of greater responsibility may appoint someone, elect someone, or inherit their status. A single individual can exercise leadership, or it can be shared by multiple people. Religious authorities like priests and secular leaders like elected councilors and village chiefs coexist in most social communities. There will be several formal leaders in any rural community, such as heads of families and kinship groups, elected councilors, heads of cooperatives, heads of religious organizations, heads of village development committees, and local leaders of political parties. Although the precise pattern will differ depending on the society, the extension agent should ascertain the responsibilities of each leader and the extent of their influence within the community. For example, a village headman might be able to provide farmers who wish to increase their holdings more land. To encourage farmers to participate in new ventures that need more land in this scenario, the extension agent will need the headman's assistance. Agents of extension should endeavor to collaborate with official leaders. They have to find out who to contact about a certain problem. Even within the vicinity of an extension agent, this could differ. In one community, an elected councilor might not have as much power as a traditional chief; in a next village, the situation might be reversed. The extension agent will not be very successful in many rural societies unless he first acquires the support of the traditional authorities. It is only at that point that he will be able to gain the community's confidence and trust. The fact that informal leaders do not occupy any specific position of authority makes them difficult to recognize. These are people who, rather than having a formal title, are admired by others for their attractive or assertive personalities, their ability to make the right decisions in any circumstance, or both. For whatever reason, they have an impact on other people. People will be more receptive to new ideas, like planting a village wood lot or establishing a cooperative, if informal leaders in the community support them. By asking farmers who they typically turn to for assistance or by keeping an eye on who stands up at village meetings, extension agents can determine the identities of these key figures. If an extension agent works through a rural society's formal and informal leaders as well as its current structure, he will be more successful. Still, there are drawbacks to this strategy. Prominent figures frequently originate from the wealthier segments of society. They can choose to retain the advantages of inputs, agricultural loans, and extension to themselves and their cronies. Extension may increase the disparity in living levels across the various societal groups by leveraging such leaders. Therefore, the agent should try to influence current formal and informal leaders, but they should also make sure that this strategy doesn't disadvantage any farmers.

1.5.1.4. Social Expectations

As was previously mentioned, a person's standing will dictate how other people view him or her. Norms are the names given to these standards. In certain societies, for instance, it is customary for a married lady to wait to eat until her husband has done. These standards have a profound effect on people's attitudes and convictions. They establish what an individual believes to be appropriate behavior in addition to how others perceive that person to behave. Regardless of how unreasonable these expectations may initially appear; extension agents should be aware of them and not undervalue their power to shape people's behavior.

1.5.2. Culture

The customs and practices that are widely accepted in a community define its culture. It is the way people live, as well as their traditions, customs, farming practices, and so forth. Every individual in a society acquires knowledge about its culture. Youngsters pick up this knowledge through observation of the behavior of older kids and adults; they are not born knowing it. As adolescents become older, they learn about the group's and society's customs and traditions from elder members of their family or kinship group. At ceremonies held later on, they might be fully initiated into society at which point they would learn about conventions, habits, and expected roles. Experience also improves one's comprehension of social norms and can impart knowledge on how to substitute more contemporary, newer behavior for some of the more conventional ones. Culture has been developed by the people to support them in their way of life; it is not merely a coincidental collection of traditions and practices. Every facet of a society's culture has a certain purpose and is, therefore, interconnected with every other facet. It's crucial to keep this in mind when organizing extension initiatives. A culture's other facets may be impacted by changes to one of its features. A program may not have much chance of success if adjustments are made to one component of culture and those changes are likely to have an unfavorable influence on other dimensions. This is one of the reasons farmers and local authorities should collaborate when creating an extension program. They will be able to determine whether society will accept the suggested modifications. An extension agent will gain greater acceptance from the people he works with if he takes the time to understand and appreciate their culture. Additionally, he will be more perceptive to the kinds of counsel and assistance that will be beneficial. The extension agent should be knowledgeable of the following five specific facets of the community's culture: the farming system, land tenure, inheritance, ceremonies, and festivals, and customary forms of communication.

1.5.2.1. Farming Systems

The extension agent needs to be aware of the farmers' current farming practices before he can advise them. Which crops are planted, and in what order or mixture? What role does each crop have in the local diet? In what ways is land ready for planting? When are the primary agricultural activities conducted? Why do individuals adopt certain farming practices? A modification in one part of a complex farming system can lead to issues in other areas. For instance, millet is occasionally sown in between maize plants in some areas of Nepal. Therefore, any modification to the spacing of maize plants or the weeding strategy that follows will have an impact on millet yield. Similar to this, up to 12 crops may be cultivated concurrently on a single plot in some parts of Nigeria. After gaining familiarity with the farming systems in the area, the extension agent can investigate potential areas for enhancement. Farmers will accept new farming techniques more readily if they can be incorporated into current systems without requiring significant adjustments. Maybe some procedures can have their timing changed, or weeding can be done more frequently. To provide more irrigated areas, alternative seed kinds may be tested, or water use could be optimized. Starting with what already exists and expanding upon it is crucial. Farming practices cannot be viewed as a merely technical subject since they are interwoven with the culture of society. It both influences and is influenced by other factors like family dynamics, food tastes, and land tenure. For instance, in one African nation, farmers were urged by extension officials to sow their crops a few weeks earlier than usual. According to research findings, farmers would have the opportunity to replant even if the early sowing failed due to a shortage of rain. Output would also increase. But this counsel called into question the legitimacy of established authorities. Until the village headman gave his approval, no one was allowed to start planting or ploughing. Additionally, the advice ran counter to the arrangement between arable farmers and cattle owners, who permitted the cattle to graze freely on the grass and stubble in the fields until planting season started. It was therefore challenging for individual farmers to modify their farming practices because this straightforward advice had ramifications for other facets of culture.

1.5.2.2. Land Tenure

How people get the authority to own and utilized property is known as land tenure. The systems of land tenure differ throughout societies. In certain societies, a tribe or kinship group owns the land, and each family is allowed to use as much of the land as they require to support themselves. There can be limitations on the purposes that land can be put to, and it cannot rent or sell the land to anybody else. In other societies, people are free to purchase land and use it whatever they like. People's ability and desire to heed recommendations to extend their stay will be impacted by the land-tenure system. For instance, share-cropping agriculture is used on some land in certain nations. The landowner receives from the farmer a set percentage of everything generated on the property. Therefore, if the landowner stands to gain the most from innovative procedures, the farmer will be reluctant to implement them. Somewhere another, the chiefs of the kinship group that owns the property forbid a young farmer from planting a tree crop. Alternatively, a tenant may opt not to fence his farm or install an irrigation pump because he fears his landlord would reclaim the land and not reimburse him for the improvements.

1.5.2.3. Inheritance

Extension work is also impacted by the transfer of land and other assets from one generation to the next. In many societies, the brothers and their offspring inherit a man's belongings instead of his own children. This could lessen the motivation of a farmer to expand the farm. It is customary in many regions for a father to leave his land to his sons and daughters before his passing. Such a farmer will not wish to alter the land in any way that would make it more difficult to cultivate each section individually in the future. Other rural societies do not pass down land through inheritance. When farmers pass away, their kinship groupings reclaim the land they farmed and reallocate it. The inheritance laws in the area should be known to extension agents, as they have an impact on young farmers' access to land and their motivation to follow their recommendations

1.5.2.4. Ceremonies and Festivals

One essential aspect of culture is rituals. They include religious festivals, times to commemorate significant seasons—like the beginning or conclusion of the growing season—and rituals for family and community life events—like marriage, childbirth, and death. In order to schedule his efforts around these, an extension agent needs to be aware of when they happen. He should also be careful to act appropriately during these situations.

1.5.2.5. Traditional Means of Communication

Every society has methods for exchanging ideas and disseminating information. In a rural location, an extension agent may encounter several traditional forms of communication such as songs, proverbs, plays, dancing, religious gatherings, and village meetings. These communication channels are crucial for extension for the following two key reasons:

1. They can tell the extension agent what the community's opinions and thoughts are. For example, knowing the meaning of local proverbs will help the agent gain insight into the attitudes of the community towards farming and their level of environmental knowledge. When organizing his programs, an extension agent should be mindful of the deeply held emotions that are frequently expressed in songs and dances.
2. These conventional channels of communication can be used by the extension agent to transmit thoughts and information. Nowadays, a lot of extension services teach new concepts using theatre, puppetry, and music.

1.5.3. Social and Cultural Change

Social structures and cultures are dynamic entities that are subject to change. The degree to which people are exposed to new ideas and cultures, as well as their own capacity to both propose and welcome change, determines how quickly society changes. The extension agent's job is to expedite cultural change in farming, even as he respects and navigates the current social structure and culture. This could then lead to more significant societal transformations. Ideas and practices gain social acceptance and eventually become considered standard. In most of what is now Botswana, land preparation was done a century and 50 years ago with hoes. An ox-drawn plough is now considered the standard tool for clearing soil and planting after farmers in what is now South Africa saw the usage of ploughs there and brought them to their own farms. Tractors are slowly replacing draft animals as a source of power in farm activities, and this has led to their cultural adoption in some areas of Pakistan and Egypt in recent times. You can also introduce new crops. Before cocoa was introduced from the United States, Ghana was unaware of its existence. Farmers in Ghana started growing it in the 1800s because traders wanted to send their cocoa beans to

Europe. Farmers acquired the knowledge and skills required to grow seedlings, ferment, and dry beans, and store them. As families relocated to different locations in order to purchase property from others on which to establish cocoa fields, land-tenure laws altered. Over time, cocoa came to play a major role in Ghana's customs, economy, and culture. In addition to being cognizant of the social and cultural shifts taking place in the region in which he operates, the agent ought to endeavor to comprehend the variables that may precipitate these shifts. Change-related factors include:

1. **Innovators:** Every culture has those members who are more open to embracing new lifestyles than others. Although these individuals do have some effect, those who are less willing to change may become suspicious of them or even jealous of them. The remainder of the community might gradually accept the new practices, though, if they are perceived as beneficial by those who have already adopted them. The inventor might then be accepted without question and possibly become more powerful. Then, general perceptions of cultural change can change; instead of being seen as a danger to long-standing practices, new ideas may be embraced as offering a better living.
2. **Contact with other cultures:** Cultural change can be attributed in large part to interactions with other communities. For instance, it was Portuguese explorers who carried cassava from South America to the west coast of Africa. Its introduction has resulted in numerous improvements to farming practices, and it is now a staple food in West African towns. In a similar vein, maize travelled to other nations with people from the United States and expanded throughout the world. Extension personnel frequently take research trips outside of their local communities. When individuals go to study or work in a different society, they often return with ideas that have the potential to alter their own way of life and influence those back home. Visitors and repeat travelers bring new fashion trends, musical tastes, political ideologies, architectural concepts, and more from one culture to the next. It is more likely that change will be welcomed by society at large the more people are exposed to new concepts.
3. **Communication:** Intercultural communication is significantly more common than it formerly was. Modern communication techniques make it relatively easy for societies to interact with one another. More locally, rural society has seen numcrous transformations because of roads and railroads. More people can visit other locations and discover new methods of doing things thanks to advancements

in travel convenience. Farmers may be encouraged to produce more in order to purchase the goods that traders place in their stores. More swiftly and affordably, farming inputs can be delivered to rural areas, and crops may be marketed with more ease. Air travel has had significant impacts as well. Air services have made it possible for remote mountain tribes in Papua New Guinea to sell produce in previously unreachable towns and mining areas. Additionally, the airplane has made parts of the high regions of Bolivia and Peru formerly inaccessible. Villagers can now receive guests from around the globe and visit other settlements. Rural residents in isolated locations can also be connected to the outside world through newspapers, radio, and television. Individuals in remote villages with access to radios or newspapers are typically well-connected and have the ability to share fresh ideas and information with their neighbors. Education is an additional means of acquainting individuals with the concepts, principles, and lifestyles of diverse nations.

4. **Population growth:** Farming systems, other facets of culture, and population size are closely related. Farmers may leave their crops after two or three seasons and move on to new, rich land in areas with low population density and abundance of farming land. Then, during the fallow season, the old fields have an opportunity to recuperate. When new ground is cleared and ready for farming, entire villages may relocate, but as the population increases, land becomes more limited. It is necessary to create new farming techniques that enable lands to be farmed year after year. Villages grow into long-term communities. This allows for the construction of more ornate homes because they won't need to be moved or abandoned every few years. Families or individuals may relocate to other locations or towns in search of employment as land becomes more and more scarce.

5. **Economic factors:** Numerous facets of peoples' lives and cultures shift as a result of economic progress. New types of work have been produced in new locations by the expansion of towns and cities, mines, and industries. To find jobs, many move from their rural homes. Many men in southern Africa leave their wives to tend to their farms for a year or so to work in the mines and cities. Women are now required to perform tasks on the farm that were formerly performed by men. In other places, on the outskirts of the city, farming might only be a part-time job. Even though most families work in cities to make their primary income, they hold onto their

farms for food and as a safety net against joblessness. Extension will be impacted by the huge number of part-time farmers. Daytime meetings might not draw much interest, and part-time farmers might not be up for new farming techniques that boost productivity if they need to spend more time in the fields. The pattern of farming is not the only part of culture that is impacted by town growth. Since individuals are no longer dependent on farms for their livelihood, laws governing inheritance and land ownership may alter. Many villagers who work in towns may not be able to attend customary rural festivals and rites, which could cause them to lose significance. Simultaneously, people who work in towns return to the hamlet with new things and money. These can raise the standard of living in rural areas and have a big impact on village values.

1.5.4. Social and Cultural Barriers to Agricultural Change

Social structures and cultures are continually evolving, although the process is frequently sluggish. There will be aspects of culture and society that could serve as obstacles to agricultural reform in the near future. It is imperative that the extension agent acknowledges the presence of these obstacles and incorporates them into their duties.

1. **Respect for Tradition:** New techniques are viewed with distrust and occasionally indifference in many rural civilizations. Respect for elders frequently leads to the belief that traditional methods are superior. Farmers are afraid of criticism for deviating from the norm in addition to the unfamiliar and untested. In these circumstances, it's common to misinterpret the intentions of change agents and other advocates. The villagers might believe that the extension agent is making modifications for personal gain. Farmers who appear to agree that a new method is desirable but are not ready to implement it can be explained by such attitudes.
2. **Belief in One's Own Culture:** Every society's member thinks their way of life is the finest. "These new methods of farming may be all right for some people, but they are no good for us." This mindset makes people reluctant to explore new things. Extension agents may encounter responses such as "We know what is best for us" and "How can it be better than our way?" to requests for change.
3. **Pride and Dignity:** Farmers could be too arrogant to engage in farming practices that could make their peers look down on them. They can be too arrogant, for instance, to transport cow dung to the fields. Though some successful farmers make more money than

most government employees and schoolteachers, farming is still looked down upon by many young people who are leaving school.

4. **Relative Values:** Extension personnel frequently highlight the increased production or financial return that comes with implementing new farming techniques. Farmers, however, might place a higher priority on beauty, flavor, or another aspect than production volume. It's also possible that they place such a high value on their free time that they are ill-prepared to put in more time on their farms. Despite having a significantly higher yield than native maize, several modified varieties of the crop have been rejected by small-scale farmers in several nations due to their poor flavor. Taste and quality of cooking are critical since farmers and their families must consume the food; they grow in addition to selling some of it.

5. **Responsibilities and Social Obligations:** People have duties that they are required to fulfil within a society or familial group. Individuals who shirk these obligations infuriate other people in the community. An individual's responsibilities to their family or community grow along with their money. Relatives will anticipate greater assistance from a farmer who makes more money. If the person feels that there won't be much personal benefit from changing their position, this might be a very substantial barrier to change. To get around this, nevertheless, one can hide cash, give cattle to friends, bury or bank money, or tell family members that there isn't any money to assist them. This could lead to a farmer, for instance, being unwilling to make obvious agricultural improvements like fencing, purchasing farm equipment, and other items that could give the impression to relatives that the farmer is wealthier than they are.

6. **Traditional Ceremonies:** Religious festivals, marriages, and funerals might take up so much time that farmers aren't able to produce as efficiently as they could. Because of this, the farmer is hesitant to accept new techniques that, while they would enhance income, would require more time to be spent on farm work and less time on social and ceremonial duties.

Although the extension agent must be aware of and sensitive to these possible social and cultural obstacles to change, their impact can be mitigated by carefully choosing the actions he advises farmers to take and the way he delivers the message. Gaining the backing of established community figures, for instance, could mitigate the impact of customs. Furthermore, the extension agent can boost the possibility that his recommendations will be followed by ensuring that popular food crops are included in agricultural programs and that the suggested varieties are acceptable based on flavor and cooking quality. Extension initiatives that attempt to introduce new methods should consider the potential impact on the culture and society, not only the technical outcomes of the suggested methods.

1.6. EXTENSION METHODS

Three categories have been established for the proper extension methods: bulk, group, and individual ways.

1. **Mass Media and Audio-Visual Aids:** There are three methods that fall under this category. The first is television and radio. These are the most widely used and understandable methods. People of all ages and backgrounds watch television and listen to radio programs to learn about a variety of topics and areas, but mostly they do it for entertainment. Radio and television help people learn and expand their knowledge of a variety of ideas and topics. Newspapers are the second source of information that people can access regarding a variety of topics, including politics, current events, science, and technology, the economy, and education. When people are working and don't have time to watch television, they read newspapers to learn new information. The third category is print media and audio-visual aids. These can be classified into many different categories, but the most popular one is television. Additionally, in the modern era, smartphones, and mobile phones are being used, and they have a significant impact on people's ability to learn. These aspects provide people with fresh perspectives and ideas.
2. **Group Extension Methods:** There are many different types of group extension techniques, such as field work, field jobs, district fairs, farm walks, farmer's rallies, folk media, meetings, group activities, tours that inspire, creation of participatory technology, training days and farmer field schools. In the case of group extension methods, people can learn and comprehend by participating in or simply taking a tour and observing various things at field events, rallies, fields, fairs, and so on. Interacting and communicating with others is a crucial skill. Through group meetings, conferences, seminars, and other events, people can gain a great deal of knowledge and understanding. When people want to improve their skills and talents, training programs are also thought to be vital. The information that is taught to participants in these programs is well-organized and presented in a logical way. Group extension methods help to improve understanding of the

technological aspects because they are important and can be applied to any field or area. Technology is seen as a subject that people sometimes find difficult to understand and feel vulnerable around.

3. **Individual Extension Methods:** When it comes to carrying out their jobs, those who operate independently, without collaborating or forming partnerships, use a variety of extension techniques. Technology is regarded as one of the most essential factors in this situation. What matters is that people must be aware of the adoption of technology; awareness, also known as knowledge, primarily refers to having understanding about novel concepts. It's important to gather knowledge and make decisions on whether to accept or reject new ideas, as well as to search out more information and change attitudes towards them. It is imperative to test a concept prior to its ultimate implementation into a task or project to assess its benefits and relevance. Additionally, innovative techniques should be utilized comprehensively. It's also said that field jobs are advantageous because they allow one to gain expertise through firsthand travels to various locations. When an individual is working exclusively on a project, it is crucial that he exercise resourcefulness, diligence, and caution when carrying out different responsibilities and duties.

1.7. CHALLENGES IN EXTENSION SERVICES

The use of technology, marketing, inputs, resources, money, education, training, and awareness, managerial skills, health conditions, natural resource availability, and environmental conditions are just a few of the many problems that have been faced in the provision of extension services (Berthe, 2015).

1. **Usage of Technologies:** Technology use is now commonplace in many types of businesses, industries, educational institutions, and training facilities. Without the use of technologies, it is difficult to implement any kind of jobs and functions. Technology is needed to prepare documents, reports, and articles. Technology has shown itself to be a significant means of communication. Through its use, one can gain enormous comprehension, knowledge, and awareness of a wide range of topics, domains, and industries. This is especially true in the agricultural sector, where it is claimed that the majority of people who live in rural areas work in agriculture.
2. **Marketing:** Increasing public knowledge of the creation of various goods and services is the primary goal of marketing. In addition to the process of creating and producing goods and services,

marketing, and promotion of those goods and services are also seen as crucial elements. There are several ways to market goods and services, including placing ads in periodicals, newspapers, and magazines; posting messages on bulletin boards; running television commercials; using mobile apps to promote goods and services; and so on. After watching these commercials, consumers acquire various goods and services based on their wants and needs.

3. **Inputs:** The terms contributions, efforts, involvements, participations, and ideas are used to describe inputs. Inputs are considered essential areas in the execution of tasks or functions, as they are necessary for the execution of all activities. Only when the inputs are appropriate can the desired outputs be obtained. People must contribute their creative thoughts and proposals, diligence, ingenuity, and optimism, as well as their sharing, involvement, and introduction of fresh methods and tactics, to the execution of extension services in agencies.

4. **Resources:** The terms sources, reserves, and supplies are used to describe resources. Resources are seen as an integral component of every project, be it manufacturing, production, or project execution. The two main types of resources are material and financial; productivity will increase when the latter are of appropriate quality. The materials and substances used in the production process are sometimes referred to as resources. The phrase is all-encompassing and encompasses approaches, methods, procedures, measurements, agencies, channels, courses, and instruments that are employed in different domains and sectors.

5. **Finance:** The most important areas for the performance of any business, service, or organization-wide operation are thought to be finance and monetary necessities. When people establish a business—a partnership or a sole proprietorship, for example—they must invest capital and work to turn a profit in order to grow their enterprise. In order to generate productivity and goodwill, it is essential to have enough financial resources and make investments before starting any program, project, plan, or endeavor.

6. **Education:** The importance of education is acknowledged by all segments of society, including those who belong to marginalized groups. Education helps people become more knowledgeable and aware of their surroundings. When someone has the right educational background, they can grow their talents and find work chances. Educational establishments and initiatives aid in the growth

of people's knowledge and abilities, enabling them to contribute to the betterment of their communities. When agencies begin offering extension services, people must meet certain educational requirements in order to be qualified to carry out all duties and responsibilities in an effective manner.

7. **Training and Awareness:** There are many different types of training programs for various fields. The primary goal of all training programs is to equip participants with the necessary skills to enable them to contribute to the efficient execution of all duties and functions. Prior to implementing new services, people must have a thorough understanding of a variety of topics and disciplines. The primary focus should be on the agencies' and organizations' potential benefits; the public, workforce, and individuals should all gain from agency extension services.

8. **Managerial Skills:** Using managerial abilities to plan, organize, direct, lead, and staff in an acceptable manner is crucial in the agencies. The process of planning helps people get from where they are now to where they want to be in the future. The task of organizing all work responsibilities, assignments, events, functions, presentations, and other activities in a suitable way is known as organizing. Giving employees proper guidance, administration, supervision, and management is the responsibility of the managing function of directing. When managers effectively lead, manage, and coordinate their workforce, they are performing the managerial function of leading. Recruiting employees based on their credentials, skills, and talents to enable them to carry out their job responsibilities effectively is a managerial role known as staffing. To control extension services in agencies, these managerial responsibilities must be carried out effectively.

9. **Health Conditions:** It is essential that people's health be in good condition when they are working hard on any job or assignment. Taking adequate breaks throughout work is crucial for individuals because working nonstop can cause stress and health problems. Since working in agencies and organizations typically entails long hours and arduous work, it is important for an individual to take adequate care of his health.

10. **Natural Resources and Environmental Conditions:** Many people, particularly those who live in rural or tribal communities, rely on the environment and natural resources. The majority of people in rural areas work in agriculture, and agricultural laborers are reliant

on the environment and natural resources. Farmers ought to be able to comprehend, adjust, and mitigate the increased challenges brought about by climate change. Since farmers must mostly rely on the weather to produce their crops, this area is seen as essential to their livelihood. Effective management of natural resources in rural communities is a crucial component of extension services. In order to guarantee that their families have access to food and a living wage, farmers, and agricultural laborers receive assistance in making the best use of the natural resources.

Chapter 2

APPROACHES AND METHODS OF EXTENSION

Contents

2.1. OVERVIEW

The profession of agricultural extension represents a nineteenth-century social innovation. It started to bring the advantages of education to the masses in rural areas after industrialization. Early extension initiatives were merely the selfless efforts of a few groups of enlightened individuals. Subsequent extension work has evolved throughout the past century because of several experiences in diverse sociopolitical contexts. Extension services are now provided through "pull" approaches, where extension is based on farmer needs, as opposed to "push" approaches, where research is shared through government extension officers. Another approach is called "innovation systems," where groups, such as farmer groups or cooperatives, are formed as middlemen between farmers, market suppliers and buyers, extension workers, and researchers. These strategies have had varying degrees of success; they have performed poorly in certain nations and successfully in others. The strategies that involve farmer groups and have a more advisory role for government agents seem to be the most promising. These strategies facilitate opportunities at the local level by coordinating networks between researchers, technical experts, and markets; they also provide regional coordination; they keep track of farmer needs and match them with research activity. There are two main methods of extension:

1. linear "push" based approaches to extension (researchers sharing with extension agents who then share with farmers), and
2. "pull" based approaches to agricultural development, where farmer needs drive extension and research.
3. A third approach type "innovation platforms" has become fashionable in the past 10 years. Innovation platforms recognize the role of social networks in generating increased productivity through access to knowledge, markets, technology, and training.

	"Push' based approaches	**"Pull' based approaches**	**"Innovation' based approaches**
Assumptions	Early adopters will create market competition that results in broad technology adoption	Farmers need to be able to understand how technology will work in their context before adoption	Develop networks and create new institutions that address value chains
Needs identification	Researchers identify how production could be improved	Farmer based needs analysis	Farmers, trainers, market organizations, researchers, extension agents

Locus and role of extension agent	Information provider and educator	Facilitator and educator	Knowledge, network, systems, and enterprise broker and convenor
Tools/Techniques	Supply chain analysis, field trials, train the trainer, train, and visit	Participatory needs analysis / PRA Farmer field schools Participatory research – farmer led experiments Farmer to farmer and group learning processes	Farmer groups and or cooperatives Group learning processes
Limitations	Assumes relevance of technology to farmers and ignores contextual barriers to uptake	Ineffective on its own as farmers have insufficient power to challenge institutional norms that influence value adding opportunities	Often focusses on early adopters who may narrow conversation scope to avoid losing competitive edge; Fails to address expectations of passive receipt of information

2.2. APPROACHES

The transition from government-based to pluralist extension service provision, wherein governments, input suppliers, exporters, NGOs, farmer organizations, and cooperatives are involved in extension providing, has coincided with the change in these paradigms. Feder (2011) offers several examples of how the public and private sectors might collaborate to provide extension. As a result, we have several strategies to deal with various circumstances, including project, commodity, privatized, participatory, and others. Similar to this, a variety of approaches, techniques, and media have been attempted to be described in this section. These include the individual to group and mass ways, which sprang from working experiences. To alter rural regions, extension work employs a variety of strategies. Diverse strategies have been used to address different demands, depending on the political and socioeconomic context. The following categories apply to extension approaches, depending on the type of intervention:

- Generalized extension approach.
- Commodity specialized extension approach.
- Educational institution extension approach.

- Project extension approach.
- Training and visit extension approach.
- Farming systems development extension approach.
- Privatized extension approach.
- Participatory extension approach.

2.2.1. General Extension Approach

This is the fundamental method of doing extension work, which is predicated on the availability but underutilization of locally relevant technologies. Farm practices would be enhanced if farmers were informed about this technology. To assist farmers in raising their output is the goal. Through organized extension services, the government steps in to help people access technology. Numerous field employees as well as a hierarchy of middle- and upper-level staff members are appointed. Plans are in place to provide guidance, credit, training, inputs, etc. There is a great need for field extension staff. Also, a lot of resources are needed, and most of the expenditures are borne by central governments. The rate at which recommendations are implemented and the rise in national output are used to gauge success. This strategy is seen as a top-down strategy.

1. **Project Approach:** A project's goal is to find a solution to a specific issue in a given time frame. The first requirement is the presence of an issue that needs to be resolved. This method, which frequently makes use of outside resources, focusses on a single area for a set amount of time. Therefore, it is assumed that rural people can significantly improve their lives with outside assistance. It is similar to doing a small-scale, intensive pilot test of a program for a set amount of time to determine if the results are good enough to move it up to a larger size. Demonstrating approaches and methods that could be continued and developed beyond the project duration is frequently one of its goals. It makes use of significant outside resource infusions for a few years to showcase the capabilities of novel technology. The Intensive Agricultural District Project (IADP), often referred to as the Package Project, aimed to promote research and technology dissemination to advance Indian agriculture.
2. **Training and Visit Approach:** The Training and Visit Approach was developed by Israeli scientist Daniel Benor and widely disseminated by the World Bank in underdeveloped countries worldwide. It was predicated on the idea that farmers needed to hear research-based farming messaging on a regular basis. For optimal performance, the extension agents must therefore be updated on a regular basis.

Without a strong management structure in place for visits, training, incentive, and resource availability as well as connections between various stakeholders, none of this would be feasible. This is an intense method built upon a meticulously scheduled program of farmer visits and extension agent and subject matter expert training. Research and extension are kept closely connected. Only extension agents engage in the transfer of technology. Increases in the production of specific commodities or crops are correlated with success. In actuality, the method operates with a small number of farmers referred to as contact farmers. It takes a unique strategy in selecting contact farmers who possess leadership qualities. But this is highly centralized, expensive, and focused on agriculture. Extension services do not aid farmers engaged in other related industries such as forestry, fishery, cattle, etc.

3. **Farming Systems Approach:** This kind of extension system or holistic approach at the local level is one of its main features. In order to address the many goals that farm households have, including nutrition, food security, sustainability, risk reduction, income generation, and job creation, the farming systems approach takes a holistic approach to the farm, farm household, and off-farm activities. This method considers the interdependencies between the elements that household members can control as well as how these elements interact with the physical, biological, and socioeconomic elements that are outside of their control. Research ties are essential, and locals are involved in an iterative process to develop technology locally that meets local needs. The degree to which locals embrace and keep using the program's developed technology serves as a barometer of success. The following is highlighted by the farming systems approach:

- that research and extension agendas should be determined by explicitly defined farmers' needs.
- an understanding of existing farming is must.
- inclusion of famers and their indigenous knowledge should be ensured.
- inclusion of social scientist and women scientists during investigation of problem is must.
- researchers, extension workers and farmers work in a team.

4. **Privatized Extension Approach:** Certain public sector reforms are now unavoidable due to mounting pressure on the government

from international agreements and financial restrictions. Intense work is required to improve agricultural productivity through value addition. Extension services are needed to meet the demands of the evolving situation. Because of the high staffing levels, transportation requirements, and other support services needed, extensions need significant investments. It is up to dispute whether it would be prudent to stick with the current extension scheme or go with one of the other options given the circumstances. While spreading beneficial agricultural knowledge to the public is crucial, export-oriented agricultural firms also need to put in a lot of work. Privatization of extension does not imply that the government department is shut down entirely and that major commercial companies are given control over extension. The government will still be able to influence how private organizations are run, oversee their performance, and mediate conflicts. When policies are favorable, private companies have occasionally contributed significantly to agricultural extension. When things go right, the private sector could be eager to offer top-notch services. While the impoverished may receive free or subsidized services, fees may be charged in order to benefit the resource-rich commercial growers. This may necessitate a close examination of private companies that are competent and offer high-quality services. Establishing performance standards is necessary to ensure quality control. Information on agriculture can now be provided via computers, video, and video text thanks to advancements in information technology. Communication mediated by satellites can make expert advice more accessible. Another option that has evolved for communication help is local radio. Therefore, the following are possible roles for public sector extension:

- To attend to target audiences unserved by private sector.
- To coordinate multiple extension providers.
- To serve as final reference or arbitrator in case of conflict.
- To assure accountability of both public and private extension services to the public.

This strategy assumes that because there are a lot of farmers to service, government extension work is becoming unsustainable and excessively expensive due to the complexity of today's technologies. There are fewer and less skilled extension workers. Big and commercial farmers will need more advanced technological know-how and abilities that local extension agents do not possess. Consequently, payment will need to be given to serve them credibly. Free service coupons or subsidized services may be provided to small and marginal farms. This will alter the function of public extension services and make them more accessible to locals (who cannot afford the entire cost). It will also encourage the development of a program that is more likely to be tailored to local needs and where extension agents are more responsible to local concerns. Its goal is to support farmers in their efforts to better themselves by offering guidance and knowledge. Global experiences indicate that small and marginal farmers are unable to afford extension services. Therefore, different arrangements are needed based on the clients' financial capabilities. There are numerous organizations whose services can be employed to provide extension services in conditions that change.

5. **Participatory Extension Approach:** This novel idea denotes community ownership and participation in the expansion process. It alludes to farmers' leadership and authority over the extension process. Farmers have provided feedback on the planning and execution of extension services. This new method is non-prescriptive and focused on farmers. Farmers are accorded the same priority as other professionals. This strategy encourages farmers to conduct their own trials in addition to valuing formal research and development operations. Through this strategy, farmer-to-farmer extension is also encouraged. The farmers must be the first to solve problems. Outsiders play a role in facilitation. The approach to participatory agricultural extension is still being developed. A lot of work is being done to develop a strategy that will enable and educate farmers to identify, address, and share responsibility for their concerns. entrusting farmers with the duty of initiating, carrying out, and assessing extension programs with outside assistance so that participation entails the people themselves recognizing the issues they confront, figuring out solutions, creating workable plans to reach these objectives, and executing them. The likelihood of success for solutions developed and implemented by the people in need is much higher than that of solutions imposed from the outside. People are thought to be wise enough to figure out their own difficulties. Since small and marginal farmers make up the bulk of farmers, they cannot rely on expensive technical solutions. As a result, students must organize into groups and carry out extended activities that fit them. Through the context of participatory extension workers can work with residents to facilitate development planning and activity execution. With regular monitoring and assessment of actions, this strategy guarantees that the extension response becomes community driven and helps village communities carry out their planned activities. The quantity of farmers who actively participate and

the longevity of regional extension agencies serve as indicators of success.

6. **Educational Institution Approach**: This strategy makes use of academic institutions with technical expertise and some research capacity to offer extension services to rural residents. Planning and implementation are frequently in the hands of those who set the curriculum for schools. Technical knowledge transfer is frequently prioritized. Those educators are usually in charge of planning and implementation. In addition to teaching and research, state agricultural universities in India are also responsible for extension. The objective is to make the technology developed at the university available to the public. The jurisdiction is selected and constrained. Individuals who work in teaching and research directly have the chance to introduce technology to their clients and witness the effects firsthand. This is also known as the nation's First Line extension system. Based on their experiences, scientists can enhance technologies for increased performance.

7. **Market-led Extension Approach**: It is a method by which an extension system can offer its customers end-to-end services, ranging from produce sales to production package procedures. It is possible to assist farmers in receiving fair compensation for their produce. A rising number of farmers now believe that the biggest obstacle to raising farm revenues is marketing, not output. Marketing extension has not yet gotten the attention it deserves because extension organizations have focused mostly on manufacturing strategies. Given the new WTO international trade environment and the export prospects that are available, this becomes even more significant. Currently, public extension workers lack the necessary tools to handle marketing extension. Extension services must address these problems by enhancing public agency capability, assisting the private sector in marketing extension, and utilizing media and IT extensively for the transmission of knowledge and technology. It is time to put marketing extension front and center. In fact, market demands will now need to heavily influence manufacturing. Farmers are aware of the factors that affect production, such as

i. what to produce.

ii. when to produce.

iii. how much to produce.

iv. when and where to sell.

In addition, farmers receive awareness training on customer preferences, market intelligence, value addition, processing, and market information systems. From agricultural produce production to commercialization, all of these aspects are covered by market-led extension (Swanson, 2008). Therefore, the following are crucial functions of market-led extension:

i. **Suggestions for product planning:** Product planning, which involves carefully choosing crops while keeping the market environment in mind, necessitates specialized guidance to help farmers remain competitive.

ii. **Marketing data:** Farmers require guidance on current pricing and trend projections. Information is required on post-harvest processing, storage, transportation, and the location of an acceptable market. Buyer, location, and crop specific information should all be included.

iii. **Suggestions for alternative marketing:** When there is excess inventory on the local market, farmers may want to consider storing. So, a well-thought-out marketing plan can contribute to increasing farmers' profits.

8. **Cyber Extension Approach:** According to Mishra (1999), cyber extension is the extension over cyber space, the virtual area that lies behind connected computer and telecommunication networks. In order to improve farmers,' extension workers,' research scientists,' and extension managers' access to information, the cyber extension strategy makes effective use of the internet, expert systems, information kiosks, and computer-based training systems. Mobile phones in particular, which are internet-connected technologies (ICT), are frequently seen as a way to enhance small-holder agricultural development. IT has helped farmers with:

- Raising awareness of market rates to ensure a fair price for produce.
- Sending orders for goods, especially for livestock and perishable goods.
- Providing up to date weather forecasting.
- Improving access to land records, to support credit applications.

2.3. EXTENSION METHODS

The two approaches are appropriate for various goals. It is crucial that the extension agent choose the approach that is best for the given circumstance by taking into account the variety of individual and group techniques at his disposal. Additionally, it's critical to keep in mind that extension work serves an instructional purpose and to make sure the chosen methodology fosters a deeper comprehension of the relevant technologies among farmers. The agent and farmer interact face-to-face in both group and individual extension techniques, and this interaction should be characterized by mutual respect and trust. As a result, the agent needs to exercise caution while applying group or individual extension techniques and make sure that he builds a sensitive rapport with the farmer. Today, there are many different approaches and mediums accessible. Here, an attempt has been made to address widely accessible techniques and media in relation to their use in extension work. Three categories of Extension methods exist. They are listed in the following order:

1. Individual contact methods.
2. Group extension methods.
3. Mass methods.

2.3.1. Individual Contact Methods

In both affluent and developing nations, one-on-one or face-to-face methods are likely the most widely employed extension techniques. Meeting with the farmer at home or on the farm, the extension agent provides information and guidance on topics of shared interest. The meetings are often informal and laid back, and the farmer can take advantage of the agent's one-on-one attention. One of the most crucial aspects of any extension job is undoubtedly one-on-one sessions, which are also very helpful in fostering trust between the agent and the farmer. Since learning is primarily an individual process, an extension agent's personal touch and one-on-one attention are crucial supports for farmers, even while group approaches allow the agent to reach a larger number of farmers. The extension worker's personal influence can play a crucial role in supporting a farmer during challenging decisions and in encouraging the farmer to engage in extension activities. The extension agent's advice is frequently heeded by farmers, who will also appreciate the personalized attention. There are several ways that this one-on-one interaction between the farmer and the extension worker might occur; these will all be discussed below.

2.3.1.1. Farm and House Visit

Typically accounting for more than half of the agent's extension efforts, farm visits are the most frequent means of direct communication between the agent and the farmer. Such visits require a lot of the agent's time, so it's critical to plan them thoroughly and to be clear about their goal. This allows for direct, in-person interaction between farmers and extension agents at the farmer's own farm. This approach is particularly useful for getting to know farmers and comprehending their issues. Sharing knowledge and instructional techniques is beneficial. If this approach is to be followed, it is preferable to continue visiting on a frequent basis. Farmers ought to be informed in advance of the visit. Having a system of paperwork to record the transactions completed during the visits is preferable. These records are a simple way to remember who said what during the past visit and to make future plans. Visits to farms can:

- familiarize the extension agent with the farmer and his family,
- enable him to give specific advice or information to the farmer,
- build up the agent's knowledge of the area, and of the kinds of problems which farmers face,
- permit him to explain a new recommended practice or follow up and observe results to date,
- arouse general interest among the farmers and stimulate their involvement in extension activities.

If the extension agent happens to be passing by and it is convenient for him to stop by, he will occasionally make an impromptu farm visit. While these unplanned visits might not have a clear objective, they are a helpful means of staying in touch and progressively establishing relationships with farmers. A little visit by the agent to greet the farmer and his family can go a long way towards building friendship and respect for one another. On the other hand, farm visits are typically scheduled into the agent's monthly calendar of activities and will form part of his overall work plan.

2.3.1.2. Office Call

He can anticipate that the farmer will occasionally pay him a visit at his office, much like the extension agent does. Such a visit frequently reflects the curiosity that the local farmers may have been piqued by the agent. Local farmers are more likely to see the extension agent if they have greater faith in him. These office visits give some of the benefits of a farm visit while requiring less time from the extension agent. Extension agents should urge farmers to stop in if it is convenient for them to reciprocate the visits, even if they wouldn't want to be inundated with them every day. A farmer or group of farmers is calling the extension agent at his office. In our nation, this method is not very widespread. But today, it's applicable in a lot of forward-thinking fields. When clients are motivated, office calls are most suitable. It uses extension workers' time and resources very sparingly. Office visits require preparation just like farm visits do. The agent may at least set up the extension office so that the visitor feels comfortable and understands what the office does, even though he may not know when a farmer is likely to stop by. The configuration might consist of:

1. making certain that the agent's name is visible and that the extension office is well marked,
2. possessing an easily readable notice board where essential, current information can be pinned,
3. containing a chair or two where guests can wait for their appointments,
4. putting on display any written extension materials that the visitor can read, such as newsletters, circulars, etc.

A trip to the extension office could be challenging for certain farmers. Therefore, the agent should make an effort to make the farmer feel comfortable and ask a few questions to help him describe his issues. The representative ought to be kind yet focused, making an immediate effort to ascertain the goal of the visit. The agent should politely end the discussion when he believes it has been sufficiently discussed to avoid having it go on in circles. Every time a visitor is escorted out, the agent should bid them farewell. The farmer's record card should also include a note about these office visits, along with any necessary follow-up.

2.3.1.3. Telephone Call

In developing countries with extensive telephone networks, this approach has been widely used. This made it easier for farmers to discuss their issues and potential solutions with the extension agent. To receive the best services, phone callers had to adhere to predetermined curtsies and other customs. The aim of phone calls and office visits is fairly similar. The extension agent is unlikely to speak with many, if any, of the farmers in his region over the phone. If a phone call is made, nevertheless, it will be to share particular information or advise rather than to have a lengthy conversation. For any reason, it's critical that the agent communicates clearly, records the key points that are discussed, and puts them in the farmer's file.

2.3.1.4. Informal Contacts

While an agent is in a certain location, they will always be informally interacting with people. He will certainly meet the farmers he works with on market days, at holidays, or at religious gatherings, and they will talk about their problems. By going to these gatherings, the agent can have a thorough understanding of the region in which he operates, as well as the issues facing the farmers. He can also exchange ideas and information informally.

2.3.2. Group Extension Methods

When working with farmers, the extension agent ought to think about implementing the group approach. Over the past 10 years, group utilization in extension has increased, and several new theories regarding the most efficient ways to employ groups have been proposed. For instance, group approaches were the foundation of the widely used Small Farmer Development Program (SFDP) in Southeast Asia, which has published two manuals that describe the methodology of group extension work. Additionally, work with extension groups in Brazil and Colombia has demonstrated the value of group formation for extension and the ways in which these groups might assist with extension activities throughout Latin America. Individual extension techniques have been shown to reach a restricted number of people and can be expensive in terms of time and limited resources. Additionally, there's a chance that placing too much emphasis on any one person could cause progressive farmers to receive too much attention at the expense of less fortunate farmers.

1. **Coverage:** The group approach is more economical since it provides the potential for wider extension coverage. By reaching more farmers through the group technique, the extension worker can establish contact with a greater number of farmers who have not previously interacted with extension activities.
2. **Learning Environment:** A more contemplative learning environment is provided by extension groups, where farmers can listen, talk, and determine whether or not to participate in the extension activity. An individual farmer can make decisions and choose a course of action with the group's backing. The club fosters a friendly environment, and by joining others to try out new ideas and discuss novel concepts, individual farmers can become more confident.
3. **Action:** Farmers with comparable issues are brought together through the group method. These issues frequently call for coordinated action (addressing the erosion of a hillside, for example), and a group is better suited to take such action than an individual, who might be overwhelmed by the scope of such issues.

It is vital to examine some of the most significant concerns pertaining to the extension group before delving into a detailed examination of several alternative group techniques of extension. A group of farmers must be formed, organized, and developed for the purpose of extension; these groups do not just materialize overnight. The extension agent's simple ability to organize the farmers for a certain task is insufficient. It is important for him to take some time to consider that the farmers will form a group, act as a group, and exhibit traits typical of groups. Experience in various regions of the world has demonstrated that the agent needs to be aware of the following four groups of crucial issues:

1. **Purpose:** When working with groups, the agent should be mindful of two primary goals. Initially, he ought to endeavor to foster the group, motivate its constituents to keep gathering, and establish the organization as a permanent entity. The agent will be creating a foundation from which the other group members can carry out their development work in this fashion. Second, the agent ought to make use of the group to disseminate fresh concepts, data, and expertise that will support farmers in their farming endeavors. Even though the second use of groups is more prevalent, it is crucial that the agent views the group's initial formation as an equally significant extension action.
2. **Size:** For rural extension, groups of 20 to 40 people are the ideal size. A huge gathering makes things unmanageable and might leave many farmers feeling disoriented and confused. Closer relationships, greater participation opportunities, and more chances to fortify friendships and support among members are all made possible by smaller groups. Geographical location is a typical factor that determines a group's size because membership is limited to individuals who reside in that area.
3. **Membership:** It is preferable to have groups of farmers with similar issues as the extension agent's role is to assist farmers in identifying and resolving issues. It could be challenging for the agent to bring the group together around a shared goal if they are dealing with a variety of farmer types, such as large landowners and small tenant farmers. As a result, the agent should closely monitor group membership and make an effort to ascertain that its members have similar issues and areas of interest.
4. **Agent's Relationship With Group:** The agent needs to think carefully about how he interacts with the group. Ideally, he will wish to support and strengthen the organization as it forms. It is envisaged that he will leave behind a structure that can operate with the least amount of extension support if his extended work takes him elsewhere. The agent should make an effort to ensure that the groupings are founded, whenever possible, upon pre-existing social or cultural community organizations rather than taking direct responsibility for their formation. The agent should exercise caution in all group interactions to avoid making the group overly reliant

on him or developing a structure that depends on him to survive. Rather, the agent ought to endeavor to foster a degree of self-sufficiency within the group by motivating the latter to spearhead extension endeavors and choose how best the agent might support them. Therefore, while the agent works with extension groups, he should keep these four difficulties in mind. His primary goals will be to do his work efficiently and make sure that more farmers are exposed to novel concepts and methods through the group approach. If he considered the group's growth during his extension work, group work could ultimately be even more fruitful and successful.

2.3.2.1. Group Meetings

The most popular technique for group extension is to call a meeting of the members of the organization or the local community. Even while these gatherings might seem casual, they nonetheless require meticulous planning and preparation. The community gathering or group provides an effective educational platform for farmers and agents to interact, allowing for the candid discussion and analysis of ideas.

It is likely that the agent can provide information regarding new agricultural ideas or practices, as well as government policies. He will want to share this new knowledge, ask the community for its thoughts, and win their support for the extension efforts. In fact, these community or group gatherings serve a wide variety of purposes:

1. **Information Meetings:** The agent gathers the community or group to share a specific piece of new information that he believes will be helpful to them and to solicit their opinion on.
2. **Planning Meetings:** Reviewing a specific issue, outlining several potential solutions, and selecting a path of action are the major goals.
3. **Special Interest Meetings:** Subjects of particular interest to a given set of people (such as dairy farming, bee keeping, or horticulture) are presented and debated in detail at a level appropriate for those attending general gatherings of the community. Discussions on topics of general community interest are open to all men, women, and youth in the community. Such general meetings should be held periodically to prevent any community group from feeling left out of extension operations. In any event, though, the agent should only arrange a meeting if he believes it will be beneficial. Farmers may decline to attend future meetings if they believe that their time was wasted at the last one, which would hinder the agent's ability to do their

job. When the agent makes the decision to call a meeting, he should carefully plan ahead and confirm a number of crucial details that are required to make the meeting go well. The agent should confer with community or group leaders to ascertain the fundamental goal of the gathering. The agent and local leaders can then discuss the meeting's best course of action and substance. It could be helpful to briefly state the goal in writing before determining which factors need to be taken into account. If the purpose of the meeting is to provide information, the agent needs to pick which order to present his material in and how to organize it coherently. He must also choose how to organize the meeting and start a conversation on the problems he wants to bring up if it is to be a general community meeting. The agent must determine the most suitable format and method of conducting the meeting based on its nature and objectives. After consulting with community members, the agent should evaluate the suitability of the various formats the meeting could take. The elements should ideally be combined according to the situation. Meetings with fewer attendees have a higher chance of catering to their individual needs. A small group of delegates will typically accomplish more than a huge assembly of all community members when plans or choices need to be made. It will be crucial to make the meeting accessible to as many individuals as possible on additional occasions. When decisions need to be made or specific business has to be handled, a formal meeting with a chairman, agenda, and written record of activities is appropriate. The meeting's decisions are accurately recorded so they can't be contested later, and the chairman keeps the discussion focused on the important matters. People feel more free to voice their opinions in an informal setting where the formality and structure of the meeting don't control them. Even while everyone can engage in an open, unstructured discussion, some individuals may end up controlling the conversation. A lecture or talk enables the agent (or other speaker) to provide his audience with a thorough, well-prepared message on a particular topic; for instance, a new technology can be presented in this manner, with visual aids to help demonstrate it. But keep in mind that lecturing is a really boring way to conduct meetings, so it's important to make sure attendees won't become disinterested. On the other hand, a lot of people can voice opinions and pose questions during a discussion. Conversations might follow a prearranged agenda of conversation topics, or they can be entirely free-form and structured.

2.3.2.2. Demonstrations

This is one of the earliest extension techniques, and it works equally well in a wide range of contexts to encourage individuals to do new things by inspiring confidence in them. To demonstrate is to exhibit. It is important to distinguish between a demonstration and an experiment (testing new ideas under artificial settings). Demonstration is the act of displaying tried-and-true methods to make an action or its worth apparent. Farmers are interested in learning how new ideas operate as well as how they might affect crop productivity. It is possible to accomplish both goals with a farm demonstration. An effective, hands-on demonstration is a crucial technique in extension work. Using the example is especially effective when working with farmers who find it difficult to read. These farmers will get the chance to witness firsthand the distinctions between a new agricultural practice that is advised and customary methods during a demonstration. The demonstration's simplicity and capacity to provide the farmers with tangible outcomes ought to be its strongest points. Extension agents primarily employ two kinds of demonstrations:

1. **Method Demonstration:** In essence, method demonstrations teach farmers how to perform a task. The farmer is guided step-by-step through the process of planting seeds in a row, using a mechanical duster to keep insects at bay, and topping tobacco during the method demonstration. Farmers who have already adopted the specific approach being demonstrated and now want to know how to do it themselves are likely the ones the agent will work with. The primary benefit of the technique demonstration is that it allows the extension agent to teach a broad audience basic farming skills, which increases the impact of his job. Furthermore, there's a better likelihood that farmers who can take part in the demonstration will gain from it than if they can only listen to it as a lecture. The primary drawback of a technique demonstration is that only a small number of farmers will be able to see, hear, and participate if there are too many in attendance. The agent needs to plan the demonstration appropriately, keeping in mind that it is a teaching tool. Additionally, it is essential that the demonstration be skillfully planned and executed.
2. **Result Demonstration:** A result demonstration's primary goal is to persuade nearby farmers that a certain new advice is workable in their environment. The key component of a result demonstration is comparison: compare poor seed with chosen seed, compare compost with no compost, or compare fertilizer application with no fertilizer. The proverb "seeing is believing" is timeless, yet it makes sense in the context of a result display. A farmer won't be persuaded by the agent's

advice until he has personally witnessed the effects of, say, applying fertilizer. The agent can considerably encourage farmers to undertake a new technique by providing them with measurable outcomes from the extension service. This will help to build farmers' trust in the method. The best way to compare new and conventional practices with farmers is to show them a result demonstration. Additionally, it can boost farmers' trust in ideas coming from research stations and aid to develop faith in more scientific farming practices. It provides evidence of the worth of a novel practice. Another helpful method an agent may use to build confidence with farmers in a new area is a demonstration of results. Its main drawback is that it requires a lot of extension resources because it takes a long time to mature. Should the new approach ultimately fail, for whatever reason, the results could be catastrophic. These kinds of failures, like the absence of rain, frequently occur for reasons unrelated to the agent's control. Extension activities like technique and result demonstrations call for a great deal of consideration, meticulous organization, and effective execution. Even though the two demos' goals are relatively different, they have a lot in common, thus they can be compared in terms of planning and execution. A number of important elements should be understood by the agent before he starts to plan and get ready for a demonstration. These points will direct his preparation and handling of the demonstration. Engagement. Whenever feasible, demonstrations have to be conducted on nearby farms with the involvement of farmers, as opposed to an extension plot or research station. When a new activity is demonstrated by a fellow farmer or is conducted by agents on extension land, farmers will be more confident than when it is conducted on the land of a neighbor. Local farmers will be more self-assured and open to learning if they are able to participate fully in the demonstration process.

i. **Simplicity:** Clear-cut, straightforward demonstrations of a single technique or novel concept will be significantly more successful than too ambitious and intricate demonstrations that place excessive demands on the farmer. It is preferable to take a few demos step-by-step rather than attempting to do everything at once.

ii. **Learning:** The demonstration should be conducted so that the farmers actually learn something because it is a learning environment. A demonstration is a form of classroom, and the agent needs to be aware of the needs of the space, time, tools, and teaching strategy to be used in the classroom.

iii. **Preparation:** An extension agent should never consider doing a demonstration without first doing extensive research and organizing. Giving a demonstration in a hurry could go very wrong.

2.3.2.3. Field days

Field days are typically less highly planned and more casual events that provide an opportunity to hold technique or outcome demonstrations on a little bigger scale. Often, the goal is to spark the interest of as many farmers as possible by introducing a novel concept and a novel crop. Field days can be held at government centers or experimental stations, but it is more common and financially advantageous for them to be held on local farmers' property. If the field day is hosted on a farmer's property and the farmer participates in organizing it and communicating its goals, there is a higher likelihood of creating an impression. Field days can be as modest as gatherings with a few farmers or as large as yearly occasions with hundreds of them.

There is less reason to worry about restricting the number because the goal is to provide a general introduction to a novel concept. On the field day, the extension agent's responsibilities include helping the farmer whose land it is being hosted on, providing general direction to make sure everything goes according to plan, and being accessible to answer questions.

It is definitely best to try to establish an environment where visiting farmers can inspect, ask questions, and generally learn about what is available rather than overly plan the field day.

The agent will make an effort to promote a relaxed and open atmosphere for the field day, but much planning is still required to make sure everything goes smoothly. We won't repeat the topics that the agent needs to think about here because they are fairly similar to those that are mentioned during demos. However, it might be helpful to keep a few more things in mind.

1. Keep the amount of people within the field's capacity to prevent crowding.
2. Make sure the field day activities are well-organized, with simple access and space for mobility.
3. Encourage the farmer demonstrating to take the lead as much as possible; encourage him but do not take over the field day.
4. In order to guarantee that everyone can hear, provide appropriately sized visual materials and, if needed, a loudspeaker. Verify whether other materials, such as extension literature, are available for consultation.

5. Gather all of the participants to wrap up the field day. Go over the events of the day, the key topics that were seen and discussed, and provide an update on any upcoming pertinent extension activities.

For farmers, a field day provides a much-needed break from their everyday labors. In order to put the farmers at ease and pique their curiosity, the agent should set up an engaging and well-presented display, offer appropriate refreshments and rest areas, and generally create a welcoming environment.

2.3.2.4. Tours

Farmers enjoy visiting other farms to observe how they operate, what crops they raise, and what kinds of challenges the farmers there are dealing with. A tour, which consists of several field demonstrations at various locations, such as farms, can pique the attention of nearby farmers. Local farmers should be able to share ideas and experiences with other farmers and observe how they manage their land through the tour. Therefore, it is critical that the agricultural practices of the area to be visited resemble those of the visiting farmers in some way. Just like any other type of extension, tours must be carefully considered, organized, arranged, and carried out. The extension agent will be guided by the five steps of identifying the goal, organizing the material, making the necessary preparations, leading the tour, and setting up the proper follow-up. But it could be helpful to include these points as well: To familiarize yourself with the local conditions, the farms to be visited, the route, and the state of the roads, first visit the area.

1. Tour only as far as it is feasible. A quick tour that allows guests to have a thorough look at nearby farms is preferable to a planned, time-constrained excursion. Avoid making the guests too tired.
2. Urge the host farmers to lead the trip and handle all of the explanations.
3. Make plans for refreshments and meals during the tour.
4. Give a brief rundown of the tour's highlights and record any observations or conclusions.

An excellent way to involve farmers and generate real interest in extension efforts is through field tours. It is also very helpful in bringing farmers together to talk about shared issues and learn from each other's valuable experiences in various fields.

2.3.2.5. Extension Talk

An extension worker's speech or speak to a small or large group of people is a very frequent way to disseminate knowledge. Planning is necessary for a formal presentation that imparts knowledge or technical expertise.

2.3.2.6. Buzz Group

This method, which is a modified version of brainstorming, is designed to start conversations in small groups and provide fresh concepts and solutions. A buzz group often has five or six people in it. Within a given time frame, the group may be invited to generate ideas on a limited or open-ended topic. At the conclusion of their discussions, the group usually selects a reporter to present their conclusions.

2.3.2.7. Brainstorming

Brainstorming is a technique for unrestricted idea generation in small groups. When fixing an issue, it is beneficial to break down barriers to money and make use of individuals' resources and experiences. In contrast to problem-solving and group discussions, the focus is on encouraging each individual to share their views without passing judgement right away. The goal is to gain from a variety of perspectives, group thought, and the synthesis of contributions in a welcoming environment.

2.3.2.8. Exhibition

An exhibition is a planned presentation of information using a range of communication channels and techniques to pique visitors' attention and inspire them to take action. A variety of goals are served by exhibitions, as listed below.

1. The purpose of the exhibition is to spark public interest in emerging technologies.
2. It contributes to publicity and goodwill.
3. When it takes place during a local festival or fair, a sizable crowd attends.
4. Using local resources and involving farmers will generate excitement.
5. It persuades people to adopt novel behavior.

2.3.2.9. Extension Campaign

A campaign is the focused employment of various media and communication techniques for a set amount of time, centered around a theme. It's a rigorous and well-planned activity.

A campaign spreads awareness among the general public and delivers a clear message. Using many channels in a dramatic way guarantees intrigue and attention.

2.3.3. Mass Media

Mass media are those communication outlets that have the capacity to simultaneously expose a huge number of individuals to the same information. These mediums include print (posters, newspapers, leaflets), moving pictures (television, film, and video), and sound (radio, audio cassettes). The rapid and inexpensive dissemination of information to a large audience is what draws mass media to extension services. Even though creating and airing a radio program can appear expensive, when you consider the millions of listeners, the cost becomes extremely inexpensive as a means of disseminating knowledge. A farmer who listens to an hour-long radio broadcast can pay less than one-hundredth of what it would for an hour-long conversation with an extension agent. Mass media, however, is not able to fulfil every role of an extension agent. They are unable to provide one-on-one guidance and assistance, impart useful skills, or promptly respond to enquiries. Given their low cost, it makes sense to use them for the activities for which they are most appropriate. Among them are the following:

1. Generating interest in agricultural advances and raising awareness of novel concepts.
2. Providing prompt alerts regarding potential pest and disease outbreaks as well as critical guidance on appropriate course of action.
3. Enhancing the effects of extension initiatives. A small number of farmers will participate in a demonstration, but if the results are publicized in newspapers and on radio, a far larger audience will be reached.
4. Exchanging experiences with other people and groups. If a village's success in starting a local tree plantation is reported on the radio, it may inspire neighboring villages to follow suit. Farmers are frequently curious to learn about the challenges faced by other farmers and the solutions they have found.
5. Addressing enquiries and offering guidance on issues that are prevalent among many farmers.
6. Repeating or reiterating guidance and facts. It's easy to forget what you hear at a meeting or that an extension agent tells you. If the media consistently reinforces it, people will recall it more readily.
7. Utilizing a range of reliable sources for farmers. Through the media, farmers can connect with prominent political figures, agricultural experts, and successful farmers in other regions, rather than relying solely on guidance from the extension agent.

Communication in the mass media demands specialized professional abilities. It won't ever be necessary for many extension agents to create radio shows or films. Nonetheless, extension agents can help ensure that mass media is used effectively by giving media producers ideas for new extension films, photographs, newspaper stories, and recorded interviews with farmers. They can also use mass media to promote their work by using it to distribute flyers and posters, or by encouraging farmers to watch farm broadcasts. In order for mass media extension to be successful, farmers need to:

1. Have access to the medium,
2. Be Exposed to the message: they may have radios, but do they listen to farm broadcasts?
3. Pay Attention to the message: information must be attractively presented and relevant to farmers' interests,
4. Understand the message.

Messages in the media are fleeting, and viewers may only give them a fleeting glance, especially if they are instructive or educational. A lot of material will quickly be forgotten if it is included in excess. Accordingly, data disseminated by the media ought to be:

1. Simple and short.
2. Repeated, to increase understanding and help the audience to remember.
3. Structured, in a way that aids memory.
4. Coordinated with advice from extension agents and other media. It's critical that the information farmers learn from extension agents and what they see and hear in the media align.

Another crucial component of communication is dialogue. However, there is little chance for a real conversation between farmers and those who create the content when it comes to mass media. As a result, media creators are ill-equipped to assess whether their messages are being comprehended by farmers or to ascertain their specific information needs. Pretesting mass media messaging and conducting research on farmers' current knowledge, attitudes, practices, and issues related to farming themes are two ways that producers might address this issue. This implies that a draft of the message is distributed to a limited group of farmers so that any necessary changes can be made before the final draft is ready, should they encounter any issues understanding it. Extension agents can assist media producers by reporting any incomprehension of the content of mass media products and by keeping them apprised of farmers' concerns and information needs. Those who create the films and advertisements for radio shows are typically better educated than farmers and don't often interact with rural residents on a daily basis. As such, they are unable to predict with any degree of accuracy how farmers would understand the content they generate.

2.3.3.1. Radio

One very helpful mass media for extension is radio. In rural towns today, battery-operated radios are a regular sight. Across a region or nation, information can instantaneously and directly reach families. Posters, extension agents, and newspapers are not the fastest ways to disseminate urgent news or warnings. Nevertheless, a skilled presenter may make shows appear quite casual and personable, creating the idea that a specific listener is being addressed to personally, even though radio has a large audience. Radio is a highly effective medium for disseminating novel concepts to a broad audience and can be employed to promote outreach initiatives. It can also make it possible for a community or group to tell others about its experiences. However, there are a few restrictions on the application of radio in extension work. In remote locations, batteries can be costly and challenging to find, and there might not be many places to fix broken radio sets. From the perspective of the listener, radio is a rigid medium: a show is broadcast at a set time of day, and if a farmer forgets to turn it on in time, they will not get another chance to hear it. The message has vanished from the record. There is nothing to remind the farmer of the information heard after the broadcast, and he cannot pause the show and go back to a point that he did not fully understand or hear. The casual manner that most people listen to the radio is another restriction. People frequently listen while engaging in other activities, such cooking, dining, or field labor. For this reason, radio is a poor medium for conveying lengthy, intricate information. Thus, a common method in many nations is to intersperse brief updates and information about farming with musical records. Another common form of radio drama is instruction offered indirectly through a play or story. Compared to a single speaker giving a conventional speech, this can capture listeners' interest and hold it longer. Lastly, audience response is scarce, with the exception of live broadcasts where viewers can call in to ask questions or voice their opinions to the host of the show. It could be challenging to provide programming that specifically address local needs in an area with just one national radio station. Furthermore, it might not be able to accommodate differences in agricultural suggestions and practices across different regions. Nonetheless, the proliferation of regional and local radio stations in recent times does facilitate the dissemination of pertinent information locally, and it also allows extension agents to participate more closely in the creation of radio programs. If local radio stations agree to let extension agents host a weekly show, they will typically provide some basic instruction on recording and broadcasting techniques. Farmers will only be drawn to farm broadcasts if they are timely and pertinent to their issues. By providing producers with information and anecdotes, as well as inviting them to visit their region to do interviews with farmers who have successfully enhanced their operations or to cover demos, exhibits, and other extension-related events, extension agents can contribute to their appeal. Extension agents can employ the following strategies to use radio more effectively:

1. **Recording:** tape recorder recording agricultural broadcasts to be played back to farmers at a later time. The amount of farmers who hear the broadcasts could rise significantly as a result.
2. **Encouraging:** enticing farmers to tune in to broadcasts, individually or in groups. Many nations have established radio farm forums, where a group gathers on a regular basis to listen to farm broadcasts, frequently accompanied by an extension agent. Following each program, they debate the material, try their best to address each other's questions, and determine whether any more action is necessary in light of what they have heard.
3. **Stimulating:** encouraging the listening habit of farming broadcasts and the belief that information obtained from the radio will be beneficial. The extension agent can accomplish this by discussing the programs' contents with farmers he meets with and by listening to them.

2.3.3.2. Audio Cassettes

Although audio cassettes are easier to use than radio, they are not as versatile as radio as a mass medium. Since cassette recorders are less widespread than radios in rural regions, locals are less accustomed to using them as information sources. The other reason that makes radio such a fast medium is because, unlike broadcast signals, cassettes also need to be physically distributed. But agents working on numerous projects have found audio cassettes to be a helpful extension resource, especially when information is too localized to be radio broadcast. The benefits of cassettes as opposed to radio are:

1. That the tape can be stopped and replayed,
2. The listeners do not have to listen at a specific time of day, and
3. The same tape can be used over and over again, with new information being recorded and unwanted information being removed.

Data can be captured on blank cassettes in the field or on cassettes in a studio, where several copies can be produced for distribution. It has already been noted that farm radio shows could be recorded for subsequent listening. Moreover, cassettes can be used for:

1. **Updating:** updating the technical data for the extension agents. Extension professionals can stay up to date on the latest technical advancements in agriculture by using the organization's pre-recorded cassettes.
2. **Sharing Experiences:** transferring experiences between and between communities of farmers. Interviews and remarks can be recorded in one village and replayed in another by an extension worker.
3. **Commentary:** giving a commentary to go along with slide sets and filmstrips.
4. **Stimulating:** encouraging debate in training facilities or farmer's groups by showcasing different viewpoints on a subject or via a videotaped play.

Recorders on cassettes are portable and somewhat sturdy. But they must be kept as dust-free as possible, and the recording heads must be kept clean with an appropriate cleaning agent, like white spirit.

2.3.3.3. Film

Film's primary benefit as a mass medium for extension is its visual nature, which allows viewers to see and hear the information it conveys. Providing visual aids for the audience makes it simpler to maintain their interest. Additionally, it enables the explanation of concepts that are challenging to put into words, such the color and form of an insect pest or the proper technique for moving seedlings. Furthermore, movement can be displayed in significantly more detail than can be seen when viewing a live presentation by using close-up shots and slow motion. It is possible to combine scenes from many eras and locations to convey concepts that are typically not visible up close. One effective way to illustrate the causes of erosion is to show how a hilltop devoid of trees allows rainwater to flow down the slope, causing gullies and topsoil removal. Similarly, crops in two different fields at various stages of growth might be filmed to demonstrate the advantages of routine weeding. After a movie is made, a lot of copies can be made, allowing thousands of people to watch it simultaneously. There are two formats for films: 16 mm and 8 mm. Most educational and motion picture releases use the bigger 16-mm format. Although the equipment and production costs for 8-mm films are significantly lower, 8 mm has only lately been considered viable for amateur home usage due to its relatively tiny projected screen size and somewhat poor picture quality. But as technology advances, more companies are creating 8-mm training and instructional videos. A projector designed for 16 mm films cannot display an 8 mm film, and vice versa. Regardless of the film format, the following items must be present: a projector; a screen or white wall to project the film onto; a loudspeaker for the soundtrack (unless the film does not have one, in which case the extension agent may require a microphone, amplifier, and loudspeaker to provide his own commentary); and a power source, which can be either a generator or mains electricity. If a generator is being utilized, it should be placed as far away from both the audience and the projector as feasible to prevent noise disturbances during the movie. It is impractical for the extension agent to screen films in villages without a vehicle because they require such heavy equipment. Films are more frequently taken to remote locations by mobile cinema vans, or they are screened in rural training facilities and schools with the necessary equipment. There are several potential drawbacks to using film as a rural extension media. The process of making a movie can take several months because it requires time for processing, editing, copying, and filming. Movies are also costly to produce, and they should only be made if they can be screened again over an extended period of time.

As a result, they are a poor medium for current information that quickly becomes outdated. The films screened for rural audiences are frequently produced in locations vastly different from those in which they are screened. The audience could find it challenging to connect the crops, cattle, agricultural equipment, people, and homes they see on film to their own farming practices. As a result, the information might not appear very relevant to them. Moreover, there's no chance for a conversation between the farmer and the film maker. Therefore, extension agents should, wherever feasible, preview a video, be ready to explain the material included in it to local farmers who might not be familiar with it, and be prepared to respond to any questions that farmers may have later. Ultimately, a film ends abruptly, much like radio shows, leaving no lasting documentation of what was seen or heard. A movie should only be used by an extension agent if it is appropriate for his extension program. A dairy farming documentary can offer insights on the types of machinery, breeds of cattle, and organizational structures that farmers may require. Once more, a proper film may explain the causes, impacts, and control strategies of soil erosion if an agent want to raise awareness of the dangers associated with it. The following are things an agent should consider when employing film for extension.

1. Choose films that complement the expansion initiative.
2. After consulting with local officials to determine an acceptable day and location, release the movie. Movies are best viewed in the evening; if the weather permits, they can be projected onto a school or other building's outside white wall.

3. Watch the movie ahead of time and determine whether the content has to be modified to reflect local circumstances. This can be accomplished by either addressing the crowd after the film has finished or by turning off the sound commentary and providing a spoken explanation.
4. In particular, if a technician is not going to be available, test out the equipment. For example, it's helpful to know how to change the projector's bulb because these things break regularly.
5. To assist the audience comprehend the material, apply it to their own circumstances, and retain it, have a discussion and ask questions after the movie.

2.3.3.4. Television and Video

Similar to radio, television is a media that can communicate information instantly to a large audience and mixes sound and visual. Cables, satellites, and land-based transmitters can all transmit television signals. Television transmission and sets are still mostly limited to cities in many nations, thus unless sets become more broadly accessible, there will not be much opportunity for television to spread to rural areas.

Compared to radios, television sets are significantly more expensive to purchase, maintain, and produce programs for. Group viewing followed by conversation has boosted access and impact when television has been utilized for rural extension communication. The majority of the benefits of both audio cassettes and film are combined in video. Images and sounds captured with a video camera are transferred to a magnetic tape and made instantly viewable on a monitor or screen. This makes it possible for the production crew to rerecord any unsatisfactory content. Similar to audio cassettes, the tape can be reused by removing any undesirable content. Video is a more versatile mass medium than film because it can produce video programs much faster and in larger quantities.

Additionally, video cassettes are lightweight and relatively simple to distribute. With the advancement of video equipment, such as television monitors and video cassette recorders, it will be feasible to deploy mobile units to present current programs produced both nationally and locally to a large number of rural people. The extension agent can clarify a point while the tape is held on a specific frame, rewind back to repeat a specific action, or slowed down. To gather content for upcoming shows, the same mobile devices might be equipped with portable video cameras. The primary restriction on watching is that whereas several hundred people may see a film projected onto a wide screen, only 20 to 30 people can watch a video program on a typical television set in a satisfactory manner. When video equipment is available, which it will be more and more in the coming years, extension agents should use the film and audio cassette usage rules provided above.

2.3.3.5. Printed Media

Printed materials can accurately and clearly communicate information by combining text, images, and diagrams. One of their greatest advantages is that they can be repeatedly referred to and viewed for as long as the viewer chooses. For this reason, they are perfect as long-term reminders of extension messages. They are only helpful, though, in places where a sizable section of the populace is literate. Newspapers, periodicals, circular letters, posters, and leaflets are examples of printed media used in extension.

1. **Posters:** Posters are helpful for informing farmers about upcoming events and for reiterating information they learn from other media. They ought to be on exhibit in busy areas that see a lot of traffic flow. The best posters are simple, grab people's attention, and are simple enough for everyone to understand.
2. **Leaflets:** Leaflets might provide precise information that is difficult to remember just by hearing, like the names of seed kinds or fertilizer application rates, or they can summaries the key ideas of a presentation or demonstration.
3. **Circular letters:** Circular letters are intended to tell farmers about local extension programs, provide timely information on issues facing farms locally, and summaries demonstration findings so that the many farmers who are unable to attend can still benefit.
4. **Newspapers:** In rural places, newspapers are not readily available. On the other hand, local officials frequently peruse newspapers; therefore, a regular column discussing agricultural matters can help raise awareness of novel concepts and educate readers about the efforts of other communities or groups.

Printed media can be produced simply and inexpensively with equipment found in many local extension offices, such as a typewriter, stencils, a duplicator, and a photocopier, or very sophisticated, with color photographs and a variety of lettering styles, requiring expensive equipment that is only available in large cities. Extension professionals may now swiftly create and distribute area-specific pamphlets and circular letters to farmers thanks to this more straightforward technology. Using two duplicators (one for black ink and one for red), leaflets can be made that are really eye-catching. Photographs cannot be replicated with stencil duplicators; therefore, illustrations must be restricted to basic outline drawings and diagrams. However, black-and-white photocopiers nowadays can make decent reproductions of these images. When an extension agent uses mass-produced printed materials, he should ensure that they support his extension efforts. For example, posters can be used to highlight a topic linked to a later demonstration; however, printed materials that the farmer does not find pertinent to the actions or statements of the extension agent will not be very effective. If printed materials are not dispersed, they are not very useful. Rather of being left to collect dust on the shelves of extension offices, expensively printed posters, pamphlets, and periodicals ought to be made widely accessible and farmers should be urged to peruse and discuss them. It's important to frequently swap out old posters for fresh ones. Furthermore, the people who created the printed materials should be informed if they turn out to be incomprehensible or irrelevant for farmers, in order to allow for necessary adjustments. It's possible that farmers don't fully understand posters and pamphlets that make sense to the extension agent. The agent should assist in clarifying their meaning wherever possible. Farmers will eventually get accustomed to the ways that words and images communicate information, and they will find it easier and easier to understand printed materials. The following steps provide a very helpful guide when the agent is developing his own printed media or when content is being created to his requirements. Posters, pamphlets, circular letters, and newspaper articles are all subject to the same rules.

1. **Define the Context:** The objective of the item should be made clear by the agency. Is the goal to raise awareness and encourage individuals to look for more specific information? or to jog farmers' memories of their acquired knowledge? or to offer thorough technical details and act as a future reference? The agency must also be aware of the audience's intended usage of the content. Will people casually walk past a notice-board and observe it? Will it be addressed in a group setting or studied on an individual basis at home?
2. **Know the Audience:** The agent must gather information on the target audience's agricultural habits, knowledge, and attitudes regarding the material's subject matter before beginning to plan the content.
3. **Decide on Content:** The information must be pertinent to farmers' requirements and appropriate in terms of both quantity and content for the media environment in which it will be used. For instance, a poster should have a single, clear message written in large, legible type that is easy for onlookers to understand.
4. **Attract Attention:** The content needs to look good right away. A person won't take the time to look at, read, and take in the information

in a leaflet or poster unless it captures his attention. Short, clearly defined headlines, striking images, and just enough white space to keep things from appearing overly crowded or packed can all assist with this.

5. **Structure the Information:** By organizing the content into logically connected sections and emphasizing the key elements with headers and highlighting, the agent can aid farmers in comprehending and remembering the material.

6. **Pre-Test:** Prior to usage, all domestically made material needs to undergo pre-testing. A few members of the target group can see it, and thereafter, questions about what they took away from it should be posed. This allows for the chance to make any necessary improvements to the material before starting final manufacturing.

2.3.3.6. Exhibits and Displays

An appealing, well-organized display not only serves as a practical means of information exchange, but it also conveys to onlookers that the extension agent and his company are effective and eager to interact. Notice boards both inside and outside extension offices, demonstration plots (where the demonstration's progress can be documented through photographs), and agricultural exhibits are good venues for displays. A well-made presentation may require a significant amount of preparation time, but a large audience will witness it. It's critical that content on permanent notice-boards is updated frequently to encourage people to check them out for the most recent information. A display ought to adhere to a particular subject that is condensed into a limited amount of messages. It should have multiple, well captioned pictures—preferably photographs—and diagrams. A large amount of printed text that isn't interspersed with images will make the display appear uninteresting and fail to draw viewers in.

2.3.3.7. Campaigns

A certain extension target is attained by utilizing a variety of media in a coordinated manner over a constrained amount of time in an extension campaign. Campaigns have the benefit of allowing the media to reinforce and support one another. The drawback is that planning a campaign might need a significant amount of time and energy. The extension agent frequently participates in national or regional staff-planned campaigns. His responsibilities will include making local arrangements for conferences, movies, press conferences, demonstrations, lodging for visiting employees, and the distribution of printed materials. A local campaign can be independently planned by an extension agent. When farmers in a region face a shared issue for which there is an easily implementable remedy, a campaign may be helpful. Careful planning is necessary for campaigns in order to maximize the utilization of all available media and extension options. Campaign planning should follow extension planning rules and principles for different media and tactics.

2.3.3.8. Traditional Media

Conventional entertainment mediums might also serve as extension media. Plays, dances, and songs can all effectively communicate knowledge. Even when they are planned ahead, they might be changed at the last minute to accommodate local circumstances and audience reaction. Modern technology is not necessary, and these materials are especially helpful in low-literate areas. Extension agents can encourage the process of problem analysis, which is a crucial component of the educational component of extension, by having locals help write the storyline of a play.

2.3.4. New Communication Technologies

Microcomputers are just one example of the new communication technologies that have arisen. Other examples include cable TV, teleconferencing, teletext, and desktop publishing. Several of these technologies, such as interactive video, compact disc read-only memory, and computer-assisted instruction, use smaller versions of once bigger devices and provide interaction opportunities (CD-ROM).

1. **Micro-Computers:** Based on factors including power, speed at which instructions can be executed, memory capacity for storing data, and level of interconnectivity, computers are categorized as mainframes, minicomputers, and microcomputers. Large and expensive was the mainframe computer. Smaller computers have less processing power. Microchip technology is the foundation of microcomputers.
2. **Electronic Mail (E-Mail):** Email, often known as electronic mail, is a technology that transmits data digitally from sender to recipient. Consequently, it's an easy method of computer-mediated communication between two people. Similar to surface mail, electronic mail is sent by phone from one computer to another when a message or letter is input into a computer. Until the recipient requests for it, the message is kept there. Computer serves as post office in this instance. Email messages can take many different forms, from voice messages to text images or drawings. A user can

view their mail box via a personal computer and modem connected to a phone with an email system. Email offers the following benefits.

3. **Interactive Video:** A video system that combines computer and video is called interactive video. It makes use of a multi-media approach, including text, slides, overheads, audio, video, and stills. The user can retrieve messages in multiple formats based on their preference.

4. **Teleconferencing:** A form of interactive group communication is teleconferencing. A technology for facilitating communication between people who are spread geographically is known as teleconferencing. Computer and telecommunications advances have made it possible for.

Chapter 3

AGRICULTURAL AND RURAL EXTENSION

Contents

3.1. OVERVIEW

Research and agricultural education are two aspects of the larger knowledge system that agricultural extension functions within. Agricultural Knowledge and Information Systems for Rural Development, or AKIS/RD, is the term used by the World Bank and FAO to describe this bigger system. It is simply known as the Agricultural Knowledge System (AKS) among the OECD's member countries. Research, extension, and agricultural higher education are the three pillars of this system that others refer to as the "agricultural knowledge triangle" (Figure 3.1). They propose that since these investments are complementary, the three pillars should be planned and sequenced as a system rather than as distinct entities (Eicher, 2001). Systematic planning is also necessary to connect the institutions within the triangle with one another and with their shared clientele, the farmers. The opportunity for AKS to address broader societal challenges related to agriculture was emphasized by government officials with AKS policy responsibilities, representatives of agricultural research, education, and extension institutes, and members of the OECD at the second AKS conference in January 2001. The participants attested to the fact that AKS may be a key player in creating research, teaching, and extension/development initiatives focused on these broader societal challenges, all of which are predicted to grow (OECD, 1999). As suggested by Eicher (2001), this situation makes the planning and sequencing of AKS as a unified system even more crucial. But without any noteworthy outcomes, a lot has been written about establishing AKIS links, particularly in research and extension (Crowder and Anderson, 1997). One concept that seems promising in this era of transition is to encourage connections between AKIS systems and their clientele through funding grants that demand cross-institutional activity.

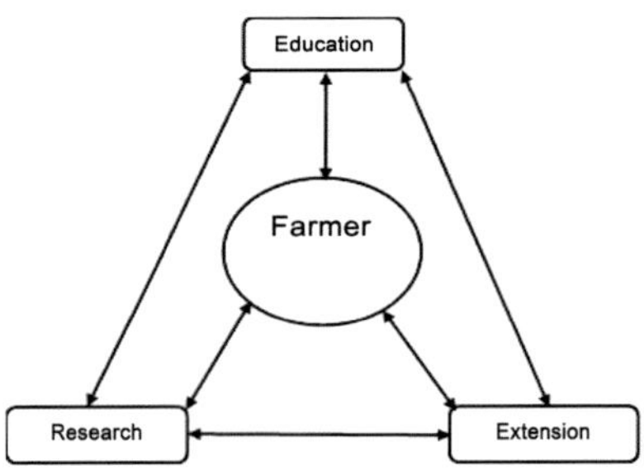

Figure 3.1. *Agricultural extension as part of AKS/AKIS.*

Source: OECD, (1999).

Agricultural information systems for rural development, as seen in the above image, connect individuals and organizations to foster learning and to create, disseminate, and utilized technology, knowledge, and information pertaining to agriculture. The system integrates farmers, agricultural educators, researchers, and extensionists, enabling them to use knowledge and information from many sources to improve farming and lives, in accordance with the AKIS/RD Strategic vision and guiding principles (FAO/World Bank, 2000). According to Maguire (2000), education for rural development and food security should replace the idea and methods of agricultural education in developing nations. In fact, a number of demands, including value-added agroprocessing, agromarket competitiveness, and trade-related education on agro-health (plant and animal health and food safety), are rising quickly. These requirements result from the responsibilities that nations bear as World Trade Organization (WTO) members and the growing need to develop competitive advantages targeted at global agricultural market niche opportunities. Basically, field observations are fed back to the agricultural education system by agricultural extension, which gets pertinent data from it.

Since agricultural vocational and higher education programs also generate the agents who work in extension, there is a professional connection between extension and these programs. Because agricultural research typically develops applied and adaptive agricultural research through development of agricultural extension, there is even tighter of a tie between agricultural research and extension. However, the topic is complicated by the fact that agricultural extension might be understood broadly or narrowly within the agricultural sector. A more expansive interpretation of agricultural extension's goals would include advancing not just production knowledge but the entire range of agricultural development tasks, including credit, supplies, marketing, and markets (agricultural process development).

Strictly speaking, the only goal of agricultural extension is to spread information to increase farmers' productivity and profitability (agricultural production performance). In its broadest sense, agricultural extension offers informal, continuing adult education related to agriculture for a variety of audiences, including farmers, spouses, youth, the community, and urban horticulturists (who seek to further their agricultural education and contribute to community development). The purposes of agricultural extension also include agricultural development, community resource development, group promotion, and cooperative organizational development. All three of the aforementioned orientations are in use in some nations, such as the United States Cooperative Extension System. These extension programs support farmers' empowerment in a number of ways, including through allowing them to participate in the design and decision-making of programs. In contrast, agricultural extension is associated with agricultural production services in numerous nations, such as Zimbabwe, Zambia, Tunisia, India, and Tunisia.

3.2. CONCEPT OF AGRICULTURAL AND RURAL EXTENSION

The objectives of the extension role are significantly broader when agricultural and rural extension are combined. For example, the Inter-American Development Bank is pushing microenterprise development, a non-agricultural activity, under the category of rural extension.

1. **Non-farm Rural Microenterprise Development:** The majority of individuals living in rural areas rely on a variety of revenue streams, including small-scale trading, primary agriculture, remittances, and temporary work. To put it briefly, rural residents do not rely exclusively on agriculture or natural resources for their means of subsistence. As Carney (1998) notes, these may serve as the foundation for their survival, but it's possible that the best opportunities for a meaningful improvement in life are found in the off-farm income generating sector, which is unrelated to the natural resources sector. The alternative of reaching the impoverished through labor-intensive rural public works projects, or rural public employment, exists in addition to microenterprise growth. Given that agriculture and rural development are combined in the AKIS/RD statement and that rural development include both farm-related and non-farm-related activities, it would seem reasonable for some extension programs to engage in activities beyond those that have already been specified. When it makes sense for extension programs to prioritize the growth of agriculture-related microenterprises in rural regions, FAO could do so by forming a specific alliance with pertinent institutions like the Inter-American Development Bank.
2. **Technical Extension:** Various technical and service units are in charge of agricultural and rural extension, which has several uses. The diverse technical units within FAO show that agricultural extension is a role that pursues a wide range of objectives, including crop improvement, animal development, fisheries engineering and capture, food, and nutrition education, and forest use and conservation. Extension may be concerned with offering information on other important topics including food storage development, processing, farm management, and marketing, even in programs intended to promote agricultural crop productivity. At one point or another, FAO has supported and worked towards all of the aforementioned goals of agricultural and rural extension.

3. **Marketing Extension:** Marketing extension is one of the other goals of agricultural and rural extension. Marketing extension (FAO, 1987) is a crucial service for nations who trade in food crops, especially perishable goods like chocolate and bananas. It offers knowledge on the post-harvest handling of specialty crops. There are other, distinct kinds of marketing information services that are also known as "market extension." These services offer data on fluctuations in commodity prices, advice on where to sell specific goods, information on issues pertaining to input quality, availability, and cost, and data on the degree of real market competition. It's important to distinguish between marketing extension services, which work to enhance the planning and execution of bringing agricultural products to market, and these market information services.

4. **Farmers' Associations:** Services for agricultural and rural extension can also assist farmers and produce processors in banding together to pursue shared agricultural goals. Group organization and promotion have a long history in extension, and FAO's dedication to these goals is well recognized. Indeed, independent agricultural and rural development group associations are one of the various ways the Organization promotes people's engagement in development (FAO, 2000). In such initiatives to build farmers' organizations, financing economic self-reliance and member engagement in the organization's activities are crucial (Rouse, 1999). Some contend that working indirectly with and via farmer groups or organizations, rather than directly interacting with individual farmers, is how extension may most successfully fulfil its mandate (Byrnes, 2001). Bingen (1998) describes how the National Union of Cotton and Food Crop Producers (Syndicat des Producteurs de Coton de Vivriers, SYCOV) came to be and emphasizes the relationship that exists between small farmer organization, democracy, and development in Cotton, Democracy, and Development in Mali.

5. **Emerging Purposes:** Extension may eventually have to deal with urban and suburban clients as populations rise and rural residents move into the metropolis (FAO, 2000). In certain cases, this is already the case. Food and agriculture, rural, and urban extension may someday replace what is currently termed agricultural and rural extension. Indeed, extension goes beyond technical agriculture and rural development to offer information and educational services in metropolitan regions in high-income countries.

6. **Urban Extension is a Potential Growth Area for Information Transfer:** As a result, it speaks to new people, offers fresh programming, and captures the rapidly urbanizing global landscape. By 2020, for example, 83% of people in Latin America would be impacted by urbanization (74% in 1998) (Sanchez-Griña, 1998). Along with epidemiological, institutional, and sociodemographic changes, this process will also impact food and nutrition through changes in socioeconomic status and demography. In North America and Western Europe, as well as in Asia and Africa, the same process is evident. Governments that are currently dismantling extension programs are likely to be forced to pay attention to issues connected to food security, youth employment in the food industry, ecologically responsible practices by small urban enterprises, and other food and agriculture-related programs. It is limiting and shortsighted to think of extension only as an agricultural output rather than an educational function.

3.3. EXTENSION APPROACHES ALTERNATIVE

In 1988, the Extension, Education, and Communication Service (SDRE) published a useful guide on alternative extension approaches, written by George Axinn, which provided an overview of several approaches and methodology. This summary, which offers a fundamental analysis of the various extension methodologies in use at the time, is still a useful reference work. The manual makes distinctions between eight methods.

When discussing extension, the terms approach and methodology are frequently used synonymously, therefore attempting to distinguish between them and their distinct conceptual implications in this study would be tedious. Let's just sum up by saying that although the phrases can be used interchangeably, they should be distinguished from the word "options," which is used in this context to refer to institutional arrangements-based reform strategies. The majority of the methods that Axinn mentions have received backing from FAO at different points in time. For the purpose of conciseness, these eight primary strategies are merely enumerated here, along with the corresponding success criteria.

1. **The general agricultural extension approach:** Success is determined by how quickly the recommendations are implemented and how much more is produced nationally.
2. **The commodity specialized approach:** Usually, the overall yield of a certain crop is used to gauge success.

3. **The training and visit approach:** The program's effectiveness is evaluated by looking at how much more of the specific crops it covers produce.
4. **The agricultural extension participatory approach:** The number of farmers who actively participate and reap the benefits, as well as the longevity of local extension organizations, serve as indicators of success.
5. **The project approach:** Short-run change is the measure of success.
6. **The farming systems development approach:** The degree to which farmers embrace and maintain the technologies created by the initiative serves as a barometer for success.
7. **The cost sharing approach:** The ability and desire of farmers to contribute to some costs, either directly or through their local government entities, is the key indicator of success.
8. **The educational institution approach:** The farming community's attendance at and involvement in the school's agricultural extension programs serves as a barometer of success.

Certainly, this is not meant to be a comprehensive list of types. More significantly, some of the approaches are distorted when Axinn characterizes them according to how successful they are. Still, it is useful to differentiate between a few fundamental strategies. Why are there so many different ways to extension? Certain concepts, paradigms, and goals are subject to change. However, teachings are also picked up and then imparted. It becomes more evident why a certain strategy has worked or not, as well as which features of a certain software are beneficial and which are not. FAO's engagement is diverse, as seen by even a quick survey of its agricultural and rural extension initiatives. It is obvious that agricultural extension uses a wide range of techniques. Additionally, it targets highly specific content areas. Additionally, it is administered and provided via a multitude of institutional frameworks. Therefore, it is reasonable to claim that, just as there is no one strategy that best suits development, no one approach best suits extended development in all situations. If not, solutions to the issues of development and extension would have been found long ago.

3.4. ROLE OF GOVERNMENT'S IN AGRICULTURAL AND RURAL EXTENSION REFORM

The government has a significant impact on rural and agricultural development, but the way it finances and delivers extensions is evolving. Government, at all levels, must be concerned with productivity, the environmental effects of agricultural methods, quality standards regulations, food safety, and the general welfare of the populace, even in cases where agricultural extension is farmer-led. A myth of the state's helplessness has emerged. It is a fact that private business extension and non-governmental organizations have frequently supplanted government extension, rendering it obsolete. Ultimately, though, the government makes the decision about whether or not to take a hands-on role in rural and agricultural extension. In addition to addressing the requirement to feed everyone, governments now have to manage natural resources responsibly, increase rural incomes, and combat poverty. The world is changing quickly, and these are important challenges.

The realities of globalization, new technologies, the emergence of new partnerships between the public and private sectors, the multidisciplinary nature of agriculture, national, and international heterogeneity, and the geographic dispersion of rural populations are all placing additional pressure on developing nations as they work towards development. Because of this, the state has to play a major role in providing funding for advising services, which are crucial but unprofitable for the private sector.

Apart from offering guidance on resource management, integrated pest control, and advisory services to the extremely impoverished, the state plays a crucial role in creating markets for commercial and farmer-to-farmer extension services, setting up rural communication infrastructure, and nurturing human capital. The proliferation of duties facing emerging countries makes the establishment of diverse alliances imperative. It is worth considering the benefits and drawbacks of institutional reform for governments that haven't done so yet. Governments and international organizations should compare and contrast the advantages and disadvantages of recently modified institutional structures for agricultural and rural extension systems in order to share knowledge and insights. Institutional improvements seem to have been implemented successfully in a number of nations, and governments may find them useful when contemplating adjustments to their own agricultural and rural extension systems. Still, no single reform initiative can be seen as a cure-all. All are ongoing projects that rely to varying degrees on the dedication, ability, resources, attitudes, and motives of the stakeholders.

3.5. GLOBAL DEVELOPMENTS SHAPING EXTENSION

The spread of new technology depends heavily on agricultural extension, yet in many African countries today, extension is either failing (as in Malawi, 2000) or dying (Eicher, 2001). Extension is either completely nonfunctional or in chaos in other low-income developing nations. The staff lacks initiative since they are unmotivated, overweight, and poorly trained. Additionally, there is minimal to no cooperation between extension and research, and even less so between extension and higher education in agriculture. Other factors, meantime, are having an impact on the growth of rural and agricultural extension services, highlighting problems and obstacles for urgently required reforms (Qamar, 2000). There is now a "complex" of extension providers that includes farmer organizations, commercial associations of extension specialists, nonprofit non-governmental organizations, and for-profit private businesses. The practice known as "contracting-in" occurs when these non-public sector extension service providers employ public sector extension agents on a temporary basis (Anderson and Crowder 2000). The realization that the world is getting bigger than a single country has a big impact on the extension of the public sector. Privatization and globalization are closely related, and nations are now dealing with a brand-new, fiercely competitive international market. Both industrialized and developing nations are undergoing significant economic restructuring, which has significantly altered the distribution of responsibilities between the public and private sectors (Fresco, 2000). Trade liberalization frequently puts developing nations at a competitive disadvantage in the world market. It begs the question, What is the role of FAO in this changing climate? Agricultural information has actually become a commodity with a price tag as it becomes more and more privatized (Rivera, 2000). The current global change of public sector agricultural extension and the development of private sector technology transfer networks are largely due to the "commodification" of agricultural knowledge. The industrialization of agriculture and the privatization of information are reflected in this shift towards the commodification of information. One outcome is that farmers are starting to pay for extension services, particularly in high- and middle-income nations. The trend shows how important agricultural knowledge is, even though this may not yet be the case in many low-income nations. Creating incomes through the information-sharing and training that agricultural and rural extension services can offer is one strategy to combat poverty. Therefore, it would seem that low-income nations have an incentive in promoting services that give their rural citizens useful, revenue-generating agricultural information. In addition to the immediate benefit of teaching their rural populations the worth of agricultural information financially, they will assist them in achieving a financial stability that will enable them to pay for these services. Modern extension reform techniques are worth considering since they can improve rural conditions in the short term by lowering poverty and have long-term effects on economic growth.

3.5.1. The New Paradigm

In many ways, the world has gotten smaller. The ease and speed at which transport and telecommunications have developed has contributed to globalization. A trend towards increased multinational business development is present. Some contend that the private sector has assumed primacy in place of the governmental sector, resulting in a "power shift." As this trend gained traction, budgetary issues in the developing nations' public sectors surfaced, and donor agencies started enforcing structural adjustment plans to align the developing nations with the financial requirements of a more stringent financial "world order." The national governments of less developed nations have been greatly impacted by these structural adjustment schemes; many have been and still are under pressure to restructure their public sector institutions. This has led to a new paradigm for market-driven reforms with an agribusiness orientation, which has a negative impact on how agricultural and rural extension are funded and delivered.

These modifications have brought about a sea change and are profoundly altering the conception and application of public sector agricultural extension. There are more competitive obstacles for the agriculture industry to overcome. People from rural areas will continue to be drawn to urban centers. Science and technological development will put more and more pressure on nations to modernize. For technologies to be effective, they need to be adjusted to new situations, and this requires a workforce that is educated. A significant investment in education, in-service job training (for undertrained agricultural extension and rural development agents, for example), and the technology system's knowledge exchange component will be necessary to prepare an educated workforce. Governments will also likely find it challenging to maintain the status quo ante. For policymakers who must simultaneously meet the demands of the global market and the requirements of their rural communities, new techniques are becoming increasingly important. If low-income nations want to survive in the global marketplace, they must unavoidably rise both qualitatively and quantitatively at their current rate of advancement. Nonetheless, it is still evident that research, extension, and agricultural education are all underfunded. Ultimately, several policy enquiries will require reexamination. These include:

1. Who will pay for such services as agricultural and rural extension?
2. Who will deliver the services? And equally important.
3. Who is to be served?
4. How will they be served? and, for what purpose?

Sometimes improving agriculture isn't the best way to solve a developmental issue. An agricultural program cannot make much of an impact and will yield little returns on investment when there is limited potential. The best answers might lie in industrialization in the long run or in out-migration in the short term. When agriculture is a feasible alternative for development, there should be a clear commitment made, acknowledging that diverse target groups would need to be catered to. Target group analysis implies that in order to satisfy a wide range of diverse objectives, distinct extension systems are unavoidably needed. Services connected to extensions will increasingly focus on meeting needs, goals, and special purposes. These services occasionally send out high-tech communications. Others will see the development of farm management abilities (Ngala, 2000). Farmers will occasionally band together for marketing objectives. To put it succinctly, a multitude of establishments and advancements will broaden the scope of organizational, technological, and educational services available for the growth of agriculture and rural areas.

3.5.1.1. Responding to the New Paradigm

At this point, farmers need to be persuaded that extension services and the knowledge they disseminate are beneficial for raising their standard of living overall and for generating money in particular. Giving resource-poor farmers access to the right technology could give rural households the chance to boost their incomes and productivity. The increased opportunities that farmers have could occasionally impede the flow of people from rural to urban areas. A large portion of the population in developing nations consists of small, low-resource farmers; increased incomes, education, and participation in development initiatives can motivate them to use land, labor, and capital resources in rural regions more effectively. Swanson (1997) states that there are various approaches to accomplish this. For example,

1. small farm households could be helped to intensify and diversify their farming systems,
2. small farmers need to be brought into the market economy,
3. small farmers also need encouragement to practice agricultural sustainability, and
4. small farmers need to be helped to organize themselves around their mutual agricultural interests.

Swanson may seem to be using top-down, supply-driven terminology, but the fundamental goals he states are in keeping with what is needed for modern growth. There is growing agreement that farmers must be directly involved in problem identification, priority setting, and on-farm research and extension operations in order to develop a demand-driven technology system (Rivera et al., 2000). In many cases, demand-driven extension is preferable, but there needs to be a balance between the government's expectations and those of farmers in various economic groups, such as estate, emergent, low-income, and marginal farmers. For nations hoping to progress to higher levels of development and competitiveness, finding a balance between farmer-initiated demand-driven extension/technology systems and institutional supply-systems should often be the ultimate objective.

3.6. FAO INPUT IN AGRICULTURAL AND RURAL EXTENSION

The development of agricultural and rural extension organizations and initiatives is greatly aided by FAO. It is devoted to promoting a variety of extension approaches and techniques, hosting international conferences, sharing knowledge with other sources, and carrying out a large number of field operations. However, FAO's particular commitment for participatory techniques and the creation of equitable processes lies beneath this variation. FAO is unique in that it places a strong focus on people-centered approaches and involvement, which is a major area of concern in its support of rural and agricultural extension. The help that FAO provides for agricultural and rural extension typically involves a variety of strategies and unique techniques for communicating with the clientele and sharing data.

3.6.1. Farmer Field Schools

An excellent FAO participatory strategy is Farmer Field Schools. Initially linked with the promotion of Integrated Pest Management, Farmer Field Schools (FFS) strive to enhance stakeholder participation in program decision-making at the local level. The ultimate goal is to assign stakeholders complete responsibility for program development. The FAO's dedication to developing participatory ways for agricultural extension is emphasized by FFS, which aligns with the organization's overarching goal of promoting equitable development. FFS began with initiatives in Asia in the middle of the 1980s and has since expanded to other continents. Furthermore, FFS is also starting to take off in Latin America as a substitute for conventional national extension initiatives in nations like Bolivia, Ecuador, and Peru. FFS is demonstrating that it is a competitive substitute for centralized, state-owned extension in its endeavors to advance farmer-led extension. The approach is presently one of the leading extension-related activities supported by FAO, and other technical services including forestry, irrigation, and water use are copying its methodology and guiding principles. Actually, an FFS project in Zambia has previously been successfully piloted by the technical section for irrigation and water use. The viewpoint offered by Quizon, Feder, and Murgai (2000) on FFS as a substitute method of problem-solving or learning is intriguing. They see FFS as an opportunity for civic engagement and empowerment rather than as an extension strategy for information dissemination. They also bring up concerns about FFS costs and their importance to the long-term viability of this strategy. The Farmers' Forest Management Schools (FFMS) are a new moniker for the FFS approach that the FAO's Forestry Policy and Institutions Branch adopted for community forestry development. Tanaka (2001) states that FFMS has two goals. First, community forest management should be made adaptable for forest users to employ for various purposes. In order for the forest to continue to sustainably supply forest users with advantages that meet their needs for a living, FFMS helps users acquire or develop the information, critical skills, and self-confidence necessary to make decisions about how best to manage the forest based on their own trials, observations, and analyses. The other is to offer a forum for discussions amongst different forest users to decide how the community forest will be used. Through the process of delegating decision-making and managing the forest, this approach also aids in the development of their sense of ownership.

3.6.2. Farming Systems Development

Farming Systems Development is another program technique that involves participation. Originally known as Farming Systems Research and Development, Farming Systems Development (FSD) changed its name to Farming Systems Research and Extension in the 1980s. A relationship between farmers, technical research, and extension is said to exist through on-farm research. This strategy has two aspects. It is sometimes praised as a collaborative, multi-institutional approach and other times as a production-oriented strategy (Berdegué, 2000). While "Farming Systems Development" (FSD) suggests more of a team-oriented approach engaging the farmers in the process, "Farming Systems Research" (FSR) tends to suggest a production-oriented strategy. According to Collins (2000), FAO has endorsed both of these system approaches. For their specific needs, FAO's technical units also maintain or make use of extension services. For example, Fisheries is developing tools for distance education to expand knowledge of contemporary fisheries engineering and fisheries capture methods. "Learning by doing" is encouraged in this computer-based remote learning program. In addition to being a significant advancement in information and communications technology (ICT), distance learning is already a top tool for knowledge and information expansion. Additionally, the livestock development unit exchanges information, particularly about services related to preventative animal health. In order to contact pastoralists and breeders, extension techniques are used. Animal health is a serious issue, as the present issues in Europe vividly demonstrate. With the growing number of services offering new technologies and specialized knowledge bases, it is becoming more and more clear why distinct information needs are being met. Even in fields where extension services are not actively created, extension is valuable and important for a variety of reasons. For example, nutrition services work in tandem with extension services to address food crises in developing nations and to oversee food delivery and education, even if they do not provide extension services in and of themselves.

3.6.3. Socio-Economic and Gender Analysis

Similar to this, extension considerations are taken into account by the Socio-economic and Gender Analysis (SEAGA) program. SEAGA, an analytical program, aims to increase awareness of gender issues and build the capacity to include gender concerns into development by providing development workers with methodologies and instruments for performing socio-economic and gender analysis. Using a bottom-up methodology to determine development goals, SEAGA prioritizes disadvantaged individuals, stresses gender roles and relationships, and encourages involvement from all stakeholders. The Integrated Support to Sustainable Development and Food Security Program, or IP for short, is an interdisciplinary program designed to create and execute an integrated approach that considers the primary social, economic, environmental, and technical facets of sustainable development. Restructuring agricultural extension systems to promote sustainable development, gender equality, and poverty reduction is one of the seven areas that IP addresses.

3.6.4. Others

In addition to the technical extension programs previously mentioned, the National Agricultural Extension Systems Reform Initiative (NAESRI) is one of the innovative programs that the Extension, Education, and Communication Service (SDRE), the lead technical unit responsible for agricultural extension and training, education, rural youth, and communication for development, has started. Strengthening decentralized extension services in the Philippines; creating a coordination mechanism for pluralistic extension delivery in Zimbabwe; conducting studies on best practices in China and Indonesia; creating participatory extension delivery strategies for female extension workers in Pakistan; determining the needs of physically disabled farmers in Iran; creating extension strategies to address HIV/AIDS in Malawi, Zambia, and Uganda; and promoting the wider adoption of hybrid rice cultivation are some of the initial projects undertaken under NAESRI. In order to assist the FAO Special Program for Food Security (SPFS) in Pakistan and Tanzania, respectively, a participatory farmer group extension technique as well as a complete extension and production support strategy have been established. Bangladesh, Egypt, Indonesia, Nepal, the Philippines, and Thailand have produced training modules on incorporating environmental and population messaging into existing extension initiatives. Guidelines for monitoring and evaluating national extension systems in the Caribbean have been developed. A number of country studies on AKIS/RD have been started in association with the Research and Technology Development Service (SDRR), the sister organization of SDRE. The Strategic Extension Campaign (SEC) approach, which has been applied in several developing nations, was also created by SDRE (Adhikarya, 1994). FarmNet is a community-based program that was started by SDRE, SDRR, and the World Agricultural Information Centre (WAICENT) with the goal of establishing farmer information networks for agricultural and rural development. In order to promote knowledge and information generation, collection, and exchange among rural residents as well as between them and the intermediary organizations that serve them, this program makes use of communication processes and instruments. One of the groups expressing interest in the program is the Uganda National Farmers' Association. Another program, VERCON (Virtual Extension and Research Communication Network), is being developed in collaboration between SDRE and WAICENT. VERCON is an institution-based initiative that seeks to facilitate communication between research and extension systems, in contrast to FarmNet. These programs aim to enhance the flow and application of agricultural knowledge in developing nations by taking advantage of the enormous potential of electronic mass media and telecommunications. A brand-new interactive initiative called "Information in Support of Sustainable Livelihoods" is being developed by WAICENT and SDRE.

This sustainable livelihood program is still in its early stages, but it aims to encourage participatory information exchange between indigenous people and the outside world through a variety of media and processes, procedures, methods, and tools. In conclusion, FAO supports the growth of agricultural and rural extension in developing nations through a multitude of diverse means. The participation of the individuals receiving assistance is a characteristic shared by these many initiatives. The aforementioned initiatives, such as FarmNet, FFMS, FFS, NAESRI, and Sustainable Livelihoods, all promote stakeholder participation in the extension decision-making processes. One of FAO's most fundamental and significant objectives continues to be empowering local people and small farmers in the use and growth of extension services through participatory methodologies. Generally speaking, the objectives of FAO's grassroots initiatives support the growth of rural populations by providing agricultural and rural extension services. Among these objectives are:

1. improving the skills of farmers, extensionists, and extension managers,
2. empowering farmers through participatory extension,
3. assisting special end-users, especially women,
4. cooperating with other agencies and organizations in extension development, and
5. assisting governments in extension policy formulation and program development.

Priority consideration should be given to enhancing the efficacy and relevance of agricultural extension programs for women farmers when developing extension institutions and programs. The diversity and continuity reaffirmed in the scope of FAO's Mid-Term Plan and executed through its technical cooperation programs are reflected in the organization's programs. The organization's ongoing initiatives lay the groundwork for future actions.

3.7. AGRICULTURAL AND RURAL EXTENSION DEVELOPMENT REFORM

Many institutional reforms have been implemented by nations all over the world in an attempt to adapt to the new paradigm. Either the market or the non-market is the focus of these reforms. This distinction, which is depicted in Figure 3.1, offers a framework for thinking about these changes as independent, non-overlapping notions that is potentially helpful but not ideal. This section is on the recent acceptance of a basic vision and guiding principles for developing agricultural knowledge systems by the World Bank and the FAO. This vision and these guiding principles are examined in light of the numerous current reforms pertaining to agricultural and rural extension. It is important to note beforehand that the kinds of extension changes that are being implemented are not always novel. Decentralization is by no means a novel concept. Decentralized authority has long existed in Australia, Brazil, Canada, Germany, India, and the United States, among other places where extension has a history of decentralization. France has been partially privatized since the 1960s. In Japan and Taiwan, participatory techniques of extension are a classic approach to the field. In Finland, pluralistic systems have a long history. However, what is novel is the degree of globalization and, as noted by Fresco (2000), the significant economic restructuring that has significantly shifted the balance between the public and private sectors in both rich and developing nations. FAO has previously implemented the reforms that are emphasized in this text. For a number of years, SDRE has been promoting changes including decentralization, participation, and provider pluralism in francophone, anglophone, and lusophone nations.

FAO's operations in the Central African Republic, Rwanda, Congo, and Guinea-Bissau are a few instances of similar initiatives in the francophone nations. Only internal technical documentation from the FAO have references to the latter attempts. Pluralism-related exploratory projects have been conducted in Zimbabwe and Mozambique. Subsidiarity has been incorporated into Uganda's present reform movement.

There are more instances of decentralization initiatives supported by SDRE in the Philippines, Indonesia, Yemen, and Iran. In fact, these countries have incorporated various components of these reforms, highlighting areas of overlap and complementarity. What is new, though, is FAO's increasing dedication to pushing these reforms as policies that other developing nations ought to take into account. The NAESRI program of SDRE provides proof of this phenomena.

3.7.1. Contemporary Institutional Reform

Extension is being shaped by global trends even more profoundly than other agricultural knowledge systems institutions. The new paradigm that supports revenue production driven by the market is one factor contributing to this drive for reform extension.

Numerous public sector reform approaches have surfaced, and they fall into various categories. Market reforms and non-market reforms are the two basic categories under which these extension reforms are arranged. This distinction indicates that market reforms include the following four main reform strategies:

1. revision of public sector extension systems,
2. pluralism,
3. cost recovery, and
4. total privatization.

Non-market reforms comprise two main reform strategies:

1. Decentralization, transferring central government authority to lower tiers of government, and
2. Subsidiarity, transferring or delegating responsibility, sometimes by abolishing authority over extension, to the lowest level of society as is practical and consistent with the overall public good (Porter, 2001).

3.7.1.1. Market Reforms

Market reforms are the outcome of the central government's goal to privatize the administration of rural and agricultural extension systems. This can be done through partnerships with farmers' associations, contracting out the provision of extension field services, or collecting costs by charging for the services. Restructuring a public sector extension system through downsizing and minimum cost recovery for services is the least drastic approach to reform it. There are several variations of this revisionist tactic. In the late 1980s, for example, the United States moved from a management-by-objectives, discipline-oriented approach to extension to an issues-oriented strategy, all the while increasingly charging for services that were previously provided for free, such attending seminars and soil monitoring. Agricultural extension units are maintained by a number of different governments, including those in Australia, Canada, Japan, Poland, Portugal, and Spain, inside their ministries of agriculture. These nations still offer extension services and finance to the public sector, either fully or partially. Nowadays, there is a lot of promotion of reforms that strengthen institutional pluralism, particularly in the developing world. This tactic frequently entails contracting out the provision of extension field services to for-profit businesses like farmers' cooperatives and consulting firms, or to non-governmental organizations like nonprofit NGOs. Contracting-out is viewed as a chance for private sector growth and public sector reform in developing nations, such as the transitional economies of Central Asia and Europe. According to Kidd et al. (2000), there have been numerous instances of the public sector working in conjunction with other organizations to privatize and commercialize the delivery of agricultural extension services. These include farmers' associations in China hiring public officials to provide technical services, profit-sharing sharecropping in Ecuador, voucher schemes in Costa Rica, subcontracting, and voucher schemes in Chile, privatized service centers in Ethiopia, contract farming in Kenya, and farmer service centers in Sri Lanka. It is important to distinguish between the institutional pluralism that the central government promotes and the reality that most nations have an institutional complex made up of private, public, and semi-public service providers (sometimes known as the "third sector" or non-profit NGOs).

When institutionalized pluralism is present, the private sector provides the extension activities, while the public sector finances them. Together with TCI and the World Bank, SDRE has been helping Mozambique plan to have non-profit NGOs provide extension services in pilot provinces under contract, while public sector delivery continues throughout the rest of the nation. In Zimbabwe, a country with a large number of active public and private service providers, SDRE recently carried out a study that may be useful in defining the function of the government in a pluralistic setting and creating a coordination mechanism. Many agricultural advice businesses in Western Europe have revamped their financial structures and are now offering fee-based or cost-recovery services to farmers. The European OECD member nations were the first to use fiscal redesign through direct charging for extension services. This was mostly due to national deficits that led to policy changes and the reorganization of their Agricultural Advisory Services. At least 20% of the OECD member countries' current budgets come from direct charges, and two of them—Norway and Finland—receive more than 50% of their funding from customers. Generally speaking, consumers of agricultural extension services are charged a fee set by the country; however, highly customized projects can fetch a premium price, while low-income users can receive a discounted rate. Some nations encourage new and low-income farmers to purchase group-use services, which are less expensive than one-on-one services, by providing discounts, subsidies, or other incentives.

In the late 1980s, Mexico became the first country in Latin America to offer fee-based services to major farmers in the northwest states that produced grain. However, Mexico ultimately decided to decentralize national extension services to the state level. China has been conducting an intriguing fee-charging extension experiment over the past 20 years (Fei and Hiroyuki, 2000). The central government of China urged local governments to increase their extension expenditures by offering fee-charging extension (FCE) services. Unlike in nations like the Netherlands or the UK, the purpose of FCE is to serve as an incentive rather than to recoup costs. By offering incentives, this system aims to motivate professionals to visit farms, stay in regular contact with farmers, and see to it that the innovations are implemented. Contract-based technology transfer is done, and it's distributed in response to farmer requests. This technique of direct contracting between the extension technician and the farmer is distinct from the schemes commonly discussed, however it is not always practical. However, as Fei and Hiroyuki (2000) point out, because the FCE model has worked well in scenarios involving small-scale farming and imperfect markets, it can serve as a valuable model for other developing nations. Nonetheless, there is a close relationship between farmers and extension technicians under this program, with rights, obligations, and financial interests linked by a direct contract between the two parties. The extension specialist must possess advanced technical knowledge and training in order to operate under such a structure. While we are talking about institutional change, it is important to keep in mind that other concerns like pre-service education and in-service training are also very important to the success of any extension program in agriculture and rural areas. Technical and administrative training is essential to the accomplishment of institutional reform. Agents must have superior technical knowledge and training in order to become extension technicians. Managerial staff must acquire superior administrative skills related to contract arrangement and monitoring, as well as supervisory abilities in human resource management and program monitoring and evaluation, in order to address the challenges posed by reform initiatives. In addition to becoming agronomic, livestock, or fisheries technicians, extension workers may also need to acquire specialized skills in farm management, farmer organization, or other programmatic and organizational abilities, depending on the objectives of the program. According to Qamar (2001), extensionists need to be ready to respond purposefully to the demands of institutional transformation. They also need to encourage the development of new knowledge, skills, and attitudes towards change. In particular, the fee-charging model is predicated on the idea that farmers will have access to useful information from skilled experts.

The complete privatization of extension, including commercialization, is one of the most dramatic current reform methods. When extension services are fully or mostly privatized, as is the situation in The Netherlands, both the financing and the provision of these services are transferred to the private sector. The Netherlands made the decision to privatize its public extension agents. Initially, this was accomplished by providing them with financial support to work with farmer associations, and more recently, DLV, a private firm, was given control over these services. Although privatization frees up funds for the government and frequently improves service performance once the private sector assumes control of the function, it may leave impoverished farmers and rural workers without any assistance. When commercialization occurs, power is delegated to a public agency that is commercialized by the government, as is the case in New Zealand. In 1986, the Ministry of Agriculture and Fisheries in New Zealand faced a challenge to transition to a user-pay commercial model and go commercial. This former government agency is now a business that goes by the name "Agriculture New Zealand." England and Wales took a similar action. In 1997, the Ministry of Agriculture, Fisheries, and Food privatized the Agricultural Development Advisory Service (ADAS), an administrative agency tasked with delivering extension services. One of the numerous private agricultural consultancies that provides primary producers in agriculture and horticulture with a variety of consulting services is this organization, which is currently called ADAS Consulting Ltd.

As was already indicated, in a number of circumstances, nations use a combination of extension structures. For example, the three categories of Norway's Agricultural Advisory Service (AAS) are based on the level of government funding. There are services that receive full government money, services that receive government funding in part, and services that receive no funding at all. Germany is an example of a combination of numerous extended decentralization systems in its three main regions: north-western, eastern, and southern (Hoffmann et al., 2000). Germany is a federally structured nation with strong states (Länder) rights. Australia has a similar situation, with each state retaining its own authority and offering different extension services, even if there seems to be a tendency towards fewer official services. To put it briefly, the reforms mentioned here are dynamic and always evolving. A number of nations, including Chile, have reviewed and adjusted their extension system arrangements multiple times. Instead of using vouchers, Chile now directly finances farmers through contracts for the use of private extension service providers. For the most part, extension reform is best characterized as ongoing effort.

3.7.1.1. Non-Market Reforms

Non-market reforms seek to absolve central government of the duty of supporting and overseeing extension, whereas market reforms aimed at the central government seek to privatize—to the fullest extent possible, either directly or indirectly—the management of agricultural and rural extension systems. The most popular non-market reform techniques are decentralization to lower levels of government, assigning non-governmental organizations the task of extension, or completely eliminating government responsibility for extension. Decentralization is frequently understood narrowly to mean extending authority to lower levels of government. For this kind of political decision, some people use the phrase devolution. Whatever the nomenclature, decentralization is a decision about policy that is distinct from market reforms. Decentralization reforms can be seen in other Latin American nations, such as Colombia, where expansion is being transferred to the municipal level, or in Mexico, where the states now have jurisdiction for the service. Some nations, like Malawi, are in the process of transitioning to a decentralized extension system at the regional level. SDRE has been working with the World Bank and TCI to help Uganda organize the transfer of fiscal and administrative responsibility to farmers' groups and local government. In a similar vein, Indonesia, Iran, and the Philippines have all received support in building and/or bolstering their decentralized extension services, and Yemen has received help in carrying out its plans to assign farmers more responsibility for extension. Economists frequently focus on the division of labor and authority between governments. They contend that the inadequacies of a feeble central government are unlikely to be remedied by the devolution of powers to local government. The impact of decentralization on research-extension links is a topic posed by Crowder (1996): does it cause these linkages to become less interconnected, or does it enhance them? On the other hand, the establishment of provincial Technology Assessment Institutes offers a new and promising paradigm of operational links among research, extension, and farmers in Indonesia, where both research and extension activities have been decentralized. These are factors that are outside the direct scope of this work, notwithstanding their importance. Additionally, governments decentralize to different levels and organizations. Currently, the development of agricultural and rural extension is dominated by at least three decentralization directions.

1. One is to reform the tax system in order to decentralize the weight of extension costs.
2. Another is to use structural reform to disperse the authority for extension within the central government.
3. The third approach involves decentralizing program administration by enabling farmers to actively participate in decision-making and, in the end, assist them in taking charge of extension programs.

Governments have the power to transfer power to farmer associations, like chambres d'agriculture or cooperatives, as is the case in Latin America's Ecuador and Europe's Denmark, Finland, and France. In certain instances, governments may transfer control to non-governmental organizations or farmer associations after relinquishing their accountability for public sector extension services. This kind of reform has been called "tertiarization" in Latin America, where non-market reforms include giving power to private NGOs, as in Bolivia, Ecuador, and Peru.

We call this final decentralization mechanism subsidiarity. The AKIS/RD Vision and Principles define subsidiarity—a relatively new term—as the idea that authority should flow to the lowest level at which all costs and benefits are internalized and economies of scale and scope are maintained. Assigning program responsibilities to local communities is a desirable approach as it involves both rural residents and local government in the planning and execution of the program.

Certain governments have transferred the institutional and technical burden of information sharing and transmission to farmers, who are now in charge of running the agricultural extension programs. It is believed that this kind of participatory involvement increases the effectiveness, sustainability, accountability, and responsiveness of services to local conditions. In Bolivia and Peru, subsidiarity takes this form.

Generally speaking, subsidiarity emphasizes the transfer of power from the government to nongovernmental organizations and, in certain situations, the complete disengagement of the government from the interests of agricultural and rural development, leaving that obligation to NGOs, farmers, and farmers' associations. The farmers' associations in a number of European nations, including Denmark, Finland, and to a lesser extent France and Germany, conduct extension with some degree of government backing. It is not the goal of this overview of institutional reforms of extension to include every reform tactic in existence, much less every structural form of extension. A number of nations have implemented reform by "deconcentrating" power to subordinate branches of the central government. This endeavor could serve as a foundation for future decentralization.

Currently, there are several ways in which central authority structures can be deconcentrated to field levels: financial grants; district administration; provincial development planning; local coordination; and regional coordination, as in Belgium, England, and Wales, Indonesia, and Ireland. Dual authority arrangements are another type of extended expansion where power is shared either

1. between the government and farmers' associations, as in Norway and Sweden, or
2. between the government and a subnational governmental entity (e.g., a state or prefecture), as in Japan, South Korea, and Taiwan.

Delegating tasks relating to agricultural development, including extension, to an outside (foreign) private enterprise has occasionally occurred. One such example is the French-run Textile Development enterprise (CFDT), which works in Africa, South Asia, and Eastern Europe. To put it briefly, nations all over the world have embraced and are still experimenting with a wide range of reform initiatives in an effort to constructively address the present paradigm shift towards democratic societies and market-oriented economies. An analysis of the diverse reform initiatives implemented globally offers governments in a transitional reform stage or reevaluating the roles of various sectors in agriculture a useful range of possibilities.

3.7.2. Dynamic View of Extension Institutional Reforms

Although this viewpoint has a few drawbacks, it aids in the clarification of the various institutional reforms related to agricultural and rural extension. Its focus on market and non-market reforms is constrained since it does not highlight potential areas of complementarity, such as connections between the development of subsidiarity within a contractual system and pluralistic changes produced under it.

As previously mentioned, another constraint is that different reform movements may exist in a given nation. In an effort to address this constraint, the following image presents the changes as though they are both part of a broader jigsaw that still revolves around factors like cost-benefit analysis, political realities, and societal requirements, and simultaneously circular in their dynamic, making them flexible (Figure 3.2).

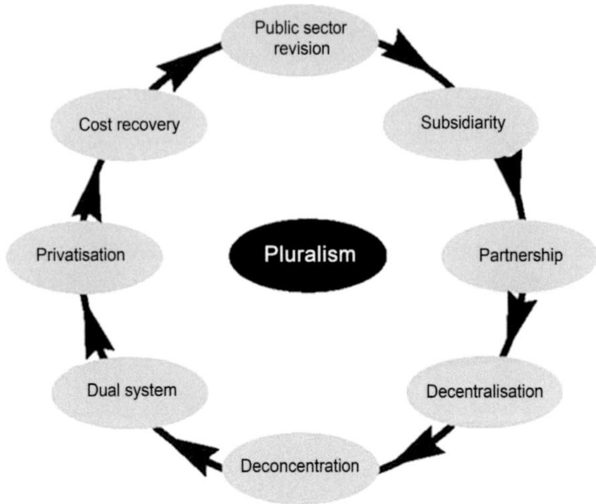

Figure 3.2. *A dynamic view of extension institutional reforms.*

Source: FAO (2001).

3.7.2.1. Principles for Extension Development

Different viewpoints on agricultural and rural extension have surfaced during the past generation. In the 1980s, a triangle with research at the top and extension and farmers at the bottom corners served as the most popular schema for expressing the relationships between extension and farming (*see* figure 3.1). The significance of research and the connections between it and farmers as well as between extension and research were emphasized by this technological triangle. Each of the triangle's three points had feedback arrows pointing from and to it. Currently, agricultural extension is seen as a component of a broader knowledge triangle that includes agricultural education, research, and extension, as previously shown in Figure 3.1. Furthermore, the phrase "Agricultural Knowledge Information System" (AKIS) is increasingly used to describe both farmers and rural communities. With this emphasis, rural residents as a whole—rather than just those involved in agricultural production—are given more attention. The AKIS/RD strategic vision still refers to the groups to be addressed as farmers, but it is not always consistent in its wider focus. Nine guiding principles serve as the foundation for the AKIS/RD vision. Among them are:

1. Economic efficiency,
2. A careful match between the comparative advantages of organizations and the functions they perform,

3. Subsidiarity,
4. Clear spread of costs,
5. Careful assessment and optimal mixing of funding and delivery mechanisms,
6. Pluralistic and participatory approaches,
7. Effective linkages among farmers, educators, researchers, extensionists, and other akis/rd stakeholders,
8. Building human and social resources, and
9. Sound monitoring and evaluation.

These lofty ideals are different in at least two areas. Many of them have to do with institutional setups that use tactics for policy reform. This is true of cost recovery, subsidiarity, and pluralism—what the AKIS statement refers to as the ideal combination of funding and delivery methods. These concepts, to put it briefly, have to do with choices for policy reform and extension finance and delivery. The remaining principles are more directly related to the development and management of programs. Issues like stakeholder involvement in decision-making, cost effectiveness, training, and development of human resources, enhancing social resources, monitoring, and evaluation are essentially program management issues and are more strongly tied to the program dimensions listed by Axinn in his approaches guide. Like other methods of extension, participation has benefits and drawbacks. It is an instrument as well as a development concept. As a philosophy, it explains the process by which all of the players work together to achieve a single objective. As a tool, it emphasizes the participation of stakeholders in activities related to decision-making, including planning, implementation, assessment, and situational analysis. The utilization of local knowledge, capacity-building, cost-effectiveness, and increased familiarity with the local environment are the benefits of the process. Therefore, one procedure that may be added to an extension system is participation. Alternatively, it might clarify the fundamental goal of the extension efforts, similar to the Farmer Field School (FFS) methodology. FFS aims to encourage rural residents to take charge of their own lives and deal with the issues, concerns, and interests that affect them. FFS is comparable to an earlier teaching/learning approach created by Brazilian educator Paolo Freire in this regard. Applying the guiding principles is the next stage in supporting extension reform, as noted in thc AKIS/RD document. The current document is meant to be part of that next stage, which is to help FAO staff operationalize the AKIS/RD vision and develop recommendations related to that vision and its principles. Case studies are being prepared in multiple countries for this exercise. Over time, a collection of protocols and resources for carrying out the suggested projects will have to be gathered and examined. But the next stage isn't as easy as picking one reform plan over another. It is necessary to comprehend the requirements and circumstances of each nation in order to analyses and discuss potential unique initiatives that may be based on but not necessarily duplicate or mimic other countries' reform efforts. The projects compiled and examined in the following part are based on the institutional reforms for agricultural and rural extension that have been, and are being, implemented globally at the start of the twenty-first century.

3.8. REFORM INITIATIVES

According to Qamar (2000), the landscape of agricultural extension is evolving. Around the world, numerous, and varied reforms have already been implemented. The decision to merge rural and agricultural extension is another example of change (FAO/World Bank, 2000). There is currently a broader perspective on agricultural extension that encompasses the growth of rural extension initiatives as well as rural development. Several tactics and approaches seem particularly acceptable when evaluating the reform strategies that are most compatible with FAO's above-mentioned goals to increase food security and relieve poverty. The three primary strategies that FAO employs are farmer group promotion, extension, and training programs, and stakeholder participation. There is considerably greater variety in the options available for advancing institutional reform. Taking into account the previously discussed market and non-market choices, and utilizing the vision and principles proposed by several international organizations, these reform approaches would consist of:

1. Encouraging pluralism and placing a focus on pluralistic alliances, such as those with private enterprise firms and farmer organizations.
2. Cost-recovery plans based on contractual clauses that shield clients from giving them unsuitable or ineffective counsel, where applicable.
3. Arrangements for decentralization with lower levels of government that grant local governments the authority to levy taxes or implement fiscal federalism.
4. Subsidiarity to farmers and farmer organizations.

The purpose of this list of strategies is to purposefully exclude those that support total privatization, in which the private sector assumes full responsibility for funding and delivering extension services. This is because it appears that government support for the development and welfare of small farmers is essential to the implementation of programs aimed at reducing poverty and ensuring food security in low-income nations. The goal behind the implementation of these and other purposeful policies must be seen with an eye towards the function of central government, not only the deconstruction or transfer of its powers, even though decentralizing and privatizing aspects are necessary. A more equal and wide-ranging group of national development actors are part of the balance-of-powers vision, which serves as the foundation upon which modern policy-driven strategies must be constructed. The long-term overarching vision and objective of reform should include the division of responsibilities, as it is necessary (Rivera, 2001). This plan calls for sharing authority between the federal government and its component parts as well as encouraging the private sector to support the growth and self-sufficiency of organized groups centered on their unique social and economic concerns. This goal would, in essence, involve promoting a balance of powers:

1. Among the various tiers of government (central, state, regional, provincial, governorate, district, and municipal),
2. Between the public and private sectors, and
3. Between government and associations, including organized citizens.

The majority of developing nations do not currently have this kind of balance of powers; hence it needs to be specifically included in the development goals agenda.

3.8.1. From Pluralism to Partnership

In nations like Uganda and Mozambique, the phrase "extension pluralism" refers to government-led development in which private extension providers receive funding to offer extension field services or are somehow integrated into the public sector extended system. Government funding for NGOs and other non-public service providers is scarce in some nations that practice extended pluralism, such Zimbabwe and Vietnam. A variety of agricultural and rural extension providers frequently operate in a nation; this is referred to as a "complex of extension-providers." It should be acknowledged, though, that pluralism is still up for discussion, just like other issues pertaining to extension and agriculture. According to Bebbington and Kopp (1998), although it did give them much-needed funding, the government's growing propensity to enter into contractual agreements with NGOs under which the latter simply carried out government programs has a tendency to erode the identity and legitimacy of these organizations. In a similar vein, Anderson, and Crowder (2000) contend that contracting-out is typically a technocratic or administrative strategy supported by funders and/or governments on a number of reasonably economic grounds. They also frequently want to maintain conceptual and methodological control, which might impede learning and adaptability. These methods, while frequently supporting several partners, don't really promote pluralistic collaborations. These remarks restate points made earlier in this paper: no single reform action can be viewed as a cure-all; rather, they are all ongoing projects that depend on the dedication, ability, capacity, attitudes, and motives of diverse stakeholders. There is undoubtedly nothing new about the agricultural extension system in a nation or region. The establishment of a diversity of agricultural and rural extension providers through public sector finance is novel. The crucial question that needs to be addressed, though, is whether or not the support of private providers by the public sector and the apparent plurality will lead to partnerships that are both meaningful and sustainable. Government-led pluralism, despite its critics, can be leveraged by developing nations to enhance the scope and efficacy of public sector extension. In addition to offering a more effective service, institutional diversity increases the likelihood that financed institutions may encourage participatory ways in extension, depending on donor engagement and conditionalities.

3.8.2. Partnerships with Farmers and the Private Sector

Each partner in a partnership has equal control over the other. Farmers' groups and decentralized government entities work together as equal partners in certain nations, such as Taiwan and South Korea. Since farmers' groups make decisions about the provision of field services, farmers in other nations (like Israel) may contract in certain services to create an equitable partnership. Other public-private partnerships that involve government support have surfaced more recently (e.g., Chile, Hungary, and Venezuela). However, the same issue about pluralism about the equity of the relationship when government pays persists when government funds commercial field service providers. A real partnership is one in which the field service is founded and runs on a 50:50 voluntary association basis with a financial budgetary partnership between the local agricultural authority and government entities, usually regional or provincial ones.

On a regional or local geographic basis, the agricultural authorities and regional councils work together to administer and run the service. The creation and delivery of field services, including budget sharing, joint central management, and field service management, typically form the foundation of such collaborations. Within the framework of a partnership, the field service conducts technological development, offers extension advice and services, and serves as the recognized authority in the technical domains and services specified in the partnership agreement. Even though there may be disagreements about the legitimacy of the more recent, contracted institutional partnerships as partnerships, they seem to be necessary to meet the increasing demand for food and maintain the base of natural resources during a period when public funding for research and extension is decreasing. According to Swanson, new institutional arrangements make it possible to work with private agricultural development companies rather than compete with them. He points out that companies in the private sector are better equipped to manufacture and market a variety of patented technologies, including improved varieties/hybrids and agrochemicals, due to their resources and comparative advantage. During this process, farmers, and eventually consumers bear the expenses of both the technical creation and transmission of these patented technologies. Swanson contends that a national technology system is more viable when the private sector is integrated with it.

3.8.3. Cost Recovery Schemes

Many approaches to cost recovery are available. There are systems where extension technicians work with farmers under a fee-based contract, and there are systems where government and private organizations charge for extension information. By requiring the stakeholders to shoulder a portion of the expense of funding extensions, cost sharing is viewed as a significant advancement that motivates them to recognize and value information. There are numerous extension cost recovery programs in the public and private sectors. Hanson and Just (2001) list the following public initiatives:

1. Fee-for-service extension provided by a public extension system,
2. Partially public-funded private extension programs, in which private companies contract to offer extension services, or in which public extension expenditures are used to pay private corporations' costs in addition to a contribution from service users, and
3. Policies that promote private extension programs, in which government mandates, subsidies, or levies on particular production techniques make it possible for private companies to offer fee-for-service extensions.

Private extension programs are distinctly private, involving fee-for-service extensions offered by private companies without government backing. Hanson and Just (2001) contend that a widespread shift to paid extension is not in the public interest, taking into account both the market failures of private extension systems, where privatization decreases social welfare provision, and the public extension failures, where privatization increases social welfare provision. They come to the conclusion that, in order to be optimal, paid extensions should be combined with public and private ones, with policy support for the latter. It is sometimes thought that charging users for advisory services will serve multiple purposes, including relieving the strain on public coffers, encouraging private sector involvement in service delivery, and holding farmers accountable as paying customers (Kidd et al., 2000). This point of view contends that the ability of farmers or their organizations to pay for services and the sustainability of agricultural markets are important factors in cost recovery. Crowder (2000) highlights that the ability and desire of farmers to pay are crucial considerations for imposing user charges on services, as demonstrated in the design of the NAADS (National Agricultural Advisory Service) initiative in Uganda. Compared to farmers with narrow profit margins and unstable markets, commercial farmers that produce high-value crops or livestock products for guaranteed markets are probably more prepared to pay for assistance. He mentions the Uganda National Farmers Union and observes that although farmers may express a willingness to pay for advisory services, the establishment of fee structures must consider both the farmers' expressed and real ability to pay. The Ugandan farmers' lack of excess financial resources prevents them from being able to pay for advising services, even in part. However, charging for an extension need not only be based on available funds; it might also be based on in-kind payments, like

1. A portion of the crop produced,
2. Services to the extension service, or
3. Selling farm-related materials.

The extension agent's advice needs to be proper in order for this to be completed. One instance of this type of fee-charging for extension can be found in China (Fei and Hiroyuki, 2000), where the farmer and the extension technician enter into contractual agreements, with payment for the extension services rendered depending on production and the sale of produce related to the farm. China's experiment is particularly intriguing because the fee-charging model serves more as an incentive mechanism than it does as a means of cost recovery. In this program, farmers, and extension technicians have a strong working relationship, with rights, obligations, and financial interests firmly entwined through a contract between the two parties. As previously stated, this kind of setup necessitates a high level of technical proficiency and training on the side of the extension technician. This system of direct contracting between the extension specialist and the farmer is different from the systems commonly discussed, but it does offer a valuable cost recovery alternative that poor countries should consider, even though it is not always practicable. Fei and Hiroyuki advocate this system.

3.8.4. Decentralization to Lower Tiers of Government

The most common way to conceptualize decentralization is as the extension of authority to lower levels of government through shifting, or devolution.

Decentralization often entails giving subnational governments more financial and managerial power. The task of formulating strategies for the reform of extension, specifically the decentralization of power to lower levels of government, has left policymakers and practitioners with a number of challenges to consider. They must determine what matters, why it matters, and how to address each one. While each country will implement reform differently, Dillinger (1995) identifies three essential components of a well-functioning decentralized government.

1. First, there needs to be a clear division of labor among the tiers of government, with strict financial limitations on the federal government's involvement in duties that are ostensibly local in nature.
2. Second, capital finance systems should prioritize creditworthiness over political savvy, and revenue sources should align with functional tasks while leaving the least amount of room for negotiation and bargaining.
3. Lastly, there needs to be an accountability framework that strikes a balance between local political involvement and federal control. In summary, Dillinger observes that financial autonomy must correspond with political autonomy and that systems based on rules perform better than those based on agreements.

Until now, the centralist legacy in emerging nations has impeded the growth of civic institutions at the community level and self-sufficiency, two essential components for the efficient operation of decentralized local administrations. The current focus on finding the best path towards the reform option of extending decentralization is a call to action for practitioners and policymakers in the sector. Holding a national workshop could be one method of creating proposals at the national level for extended decentralization.

An overview of global institutional reform initiatives and a gathering of present and future stakeholders to discuss different decentralization arrangements—including challenges and achievements in other nations—would be the goals of such a workshop. The purpose of the national workshop could be to make clear the path these stakeholders believe the nation should take in order to pursue decentralization reform of rural and agricultural extension. In order to provide a menu of alternatives when facing the challenge of decentralizing public sector agricultural extension, the various structural and financial arrangements that governments around the world have adopted in the last 20 years to improve agricultural extension services will be analyzed and compared. However, local governments need to be somewhat financially independent of the federal government for decentralization to succeed. Shah (1994) highlights the significance of fiscal federalism, which allows the subgovernment level to support its activities, including services like agricultural and rural extension, by means of block grants or the authority to levy taxes.

3.8.5. Subsidiarity (Decentralization) to the Grassroots Level

When operational authority and responsibilities are delegated to the lowest level of authority that is consistent with organizational competences and the economical use of resources, this is referred to as subsidiarity. Funds and other resources would be distributed to the grassroots level in accordance with designated roles. It could be worthwhile to investigate the degree to which governments are open to and prepared to transfer the responsibility for delivering extension services to farmers' associations and thriving rural community groups. Although subsidiarity implies that the lowest echelon of society should have the most authority, this action should be reasonable and in line with the general welfare of society. Sometimes, assigning extended tasks to grassroots organizations means making a decision between giving local government and/or community organizations more authority. Making sure that all costs and benefits are internalized and that economies of size and scope are maintained is crucial. Empowering local communities to oversee programs—as Colombia has done—is also a desirable approach since it involves rural residents in the planning and execution of the initiatives. Governments are starting to use institutional and technical mechanisms to transfer management of agricultural extension programs to farmers. Participatory involvement is thought to improve services' accountability, effectiveness, sustainability, and responsiveness to local conditions. In locations as far apart as Bolivia and Uganda, this type of subsidiarity is taking place. The government may decide to go in numerous ways when it comes to implementing grassroots command of extension. For example, the government of Uganda is attempting to give districts and sub-counties more fiscal authority while giving farmer associations the ability to contract out extension services to private companies. Together with other commercial non-governmental organizations, it is anticipated that the current public sector extension agents would establish extension consultancies and turn into for-profit suppliers.

Chapter 4

AGRICULTURAL EXTENSION: CHALLENGES AND SOLUTIONS

Contents

4.1. OVERVIEW

Today's world faces several issues, including poverty, hunger, economic growth, food supply, and the depletion of natural resources. At least 800 million people experience chronic hunger today, a situation that is likely to worsen when the world's population rises to 8000 million by 2025. Even if the population of cities triples over that time, widespread poverty will still mostly affect rural areas. Reducing poverty, facilitating food security, and implementing sustainable resource management all depend on a robust agricultural economy. Over 60% of people in developing nations make their living from agriculture, and in many of these nations, farm families account for 80% or more of the total population (World Bank, 1990). Since agriculture has already depleted water and land, any further gains in food production will need to take advantage of biological yields on currently cultivated land (World Bank, 1997). Many developing nations are experiencing a crisis in agriculture as a result of this technological problem. The green revolution's accelerated growth pace in Asia has slowed. Recurrent famine is a result of Africa's declining per capita food output, which has been the case in most years since 1970. Population pressure and intensive agriculture pose major environmental threats in many parts of Latin America and the Caribbean (World Bank, 1990). Additionally, resistance to high input agriculture is growing in the industrialized world due to concerns about genetically modified products, animal rights, and contamination of the soil and water.

4.1.1. Challenge for Extension

There are two main obstacles that agricultural extension must overcome as the globe struggles with these problems. In the agriculture industry, organization, and information must be given more weight. To establish an agriculture that will meet complicated demand patterns, decrease poverty, and conserve or enhance ecological resources, those working in the industry need to increase their skills, knowledge, and ideas. Extension has a significant part. The extension mandate presents challenges for both funding and delivery of extensions:

1. magnitude of the task,
2. dependence on wider policy and other agency functions,
3. problems establishing the cause and effect necessary to obtain political and financial support,
4. liability for public service functions beyond agricultural knowledge and information transfer,
5. fiscal sustainability, and
6. interaction with knowledge generation.

Public extension is perceived by many as not doing enough, not doing it properly, and not always relevant. Poor program design and implementation, coupled with bureaucratic inefficiencies, have resulted in poor performance and illogical connections between the research industry and clients in developing nations. The 1980s saw a fall in support for extension, and donors were reluctant to provide funding for significant recurrent expenditures in the public sector. This resulted in additional underfinancing, a shortage of staff, and a reduction of extension services (Amanor and Farrington, 1991). Policymakers have a variety of options for extensions, including priorities, goods, mandates, and models, and they must weigh conflicting opinions about what extensions should accomplish while searching for solutions. Due to budgetary constraints, there is tremendous need to show the return on investment for extension and investigate private, local government, and producer group alternatives to public finance.

4.1.2. Government Investments in Extension

Approximately 1200 million people are reached by agricultural extension globally, which employs at least 800,000 extension professionals and hundreds of thousands more farmer technicians or leader farmers.2. Approximately 80% of extension services provided worldwide are currently paid for by the public and provided by state personnel (World Bank, 1997). Roughly 12% of services are provided by universities, parastatals, and nongovernmental organizations; the remaining 5% are provided by the private sector. There are typically many different degrees and forms of government participation. It may pursue a variety of goals and fund, staff, or enable extension by creating favorable rules and guidelines for other providers. This complexity contributes to a few of the general extension-related issues. Most governments assumed exclusive responsibility for extension delivery due to the conventional understanding of the "public good" of various components of agricultural information transmission (Umali-Deininger, 1996). Social goals like reducing poverty and focusing on certain populations, such rural women, have also been used to justify public support for many extension programs (Wilson, 1991). In 1990, the World Bank stated that there was "no substitute (for a large nationwide public extension system) for a corps of well-trained and well-supported extension workers and subject matter specialists to serve as competent, motivated, and trusted "agents of change.'" Governments in developing nations made significant investments in agricultural cxtcnsion because they anticipated higher agricultural output. Real expenditure on extension increased more than sixfold in Latin America, tripled in Asia, and more than quadrupled in Africa between 1959 and 1980 (World Bank, 1990). In order to gather information about the current state of agricultural extension in 113 countries, FAO contacted 207 organizations that were thought to be broadly representative of agricultural extension systems worldwide (Swanson et al, 1984). The funding source for 86% of these organizations was a ministry of agriculture or a comparable government organization. FAO projected, based on this survey, that $6000 million was spent globally in 1988 on public agricultural extension agencies, or $8522 on average per extension worker. In low-income nations, the average annual farmer expenditure on extension services was $2–3, but in high-income countries, it was $65. The percentage of the labor force that is employed in agriculture in various nations is directly correlated with the amount of public investment in agricultural extension as a percentage of total MOA (Ministry of Agriculture) resources. Extension receives roughly 2% of ministry resources when less than 20% of the population works in agriculture, compared to slightly more than 20% of resources when more than 60% of the population works in agriculture.

Extension-related changes and difficulties are a sign of larger social factors at play. Extension-related public policy typically follows development policy, and the institutional evolution of extension mirrors developments in other institutional domains. Modifications to the economic, political, technological, social, and financial settings for extension have brought about significant trends and changes in extension praxis. The early colonial emphasis on commodity programs—many of which persist today—was replaced in developing nations by more comprehensive multipurpose rural development initiatives. These initiatives were frequently established by colonial powers, who dispatched expatriate "rural agents" to organize communities and act as points of contact for the government, input, and credit providers, and buying agents. Their widely outlined responsibilities frequently included tax collection, census taking, and human health. The 1950s saw the establishment of several national extension services, the creation of nation states, and state-led, planned development. Agricultural extension was linked to ministries of agriculture due to the lack or weakness of agricultural universities, resulting in a weak connection to research (Axinn, 1988). The "diffusion model' of MOA extended distribution, a hierarchical, unidirectional method of technology transfer supported by advancements in mass media, was made possible by confidence in Western technology. The 1960s marked the start of the green revolution as well as the period of interpersonal contact and community development. There was still a heavy emphasis on technological transfer. Multipurpose agents started to be replaced by systems that were more closely focused on extension and its management starting in the mid-1970s, when public sector extension restricted guidance to technical agricultural matters, mostly major annual crops (World Bank, 1990). According to Swanson et al. (1990), half of the extension agencies surveyed by FAO in 1989 underwent organizational changes or reforms starting in 1970. The 1970s were characterized by integrated approaches to rural development and the growth of the training and visit (T&V) extension system. The "get the technology right' approach replaced the "diffusion' model of extension, explaining non-adoption of technology through farm-level restrictions and offering integrated service packages as a remedy (Axinn, 1988). Participatory tactics became increasingly important during the transformative 1980s. Along with attempts at cost recovery and privatization, concerns about maintaining ecosystems and boosting the productivity of women were introduced (World Bank, 1990). The 1990s are now the alternative period, when innovative methods are being tested in a tight budgetary setting. Power in the public sector has dramatically decreased as a result of democratization. Methodologically, it is important for researchers and farmers to have direct ties at the farm level. Extension funding strategies that are more sustainable have more partners and more flexibility.

90% of extension workers worldwide are based in poor nations, with more than 70% of them in Asia (Umali and Schwartz, 1994). Approximately 95% of extension staff members are employed in public agricultural extension systems. In developing nations, public extension agencies' extension coverage—that is, the proportion of extension staff to farmer population—varies from 1:1800 to 1:300. The ratios in developed Asia, North America, and Europe average roughly 1:400. In many developing nations, there is a large workforce, but the quality of the workforce is frequently low. Managers of public extension systems must deal with budgetary constraints, which compel them to engage unskilled workers in order to reduce operating expenses. FAO discovered that just 40% of extension staff members had completed secondary school, and another 33% held an intermediate credential. According to FAO data, commercial farmers—including specialized producers of cash and export commodities—received about 58% of extension resources. Slightly over one-third of all extension resources go to the great majority of smaller, marginal farms. Not all extension is closely associated with the dissemination of agricultural knowledge. According to FAO research, non-educational activities account for around 114% of the time spent by extension workers in developing nations, or "about 140,000 full-time equivalent years of extension personnel time." The Food and Agriculture Organization considered this a "major loss of educational resources, especially in developing countries where extension coverage is still grossly inadequate, both in quantitative and qualitative terms." Extension agents may spend a lot of time in commodities programs assisting farmers in obtaining inputs, financing, and marketing services. However, in addition to responding to information requests from ministry headquarters or extension, general government extension staff may also be involved in a range of local government servicing activities. Roling (1986) interprets the range of extension functions in terms of two traditions:

technical innovation (TI) and human resource development (HRD). According to him, the majority of extension organizations worldwide focus solely on technology transfer (TI) and are funded by tax revenues in an effort to maximize the efficiency of food, raw material, and export commodity production. Human Resource Development (HRD) is centered on rural populations and the social systems that support them. It addresses issues including farmer organization and mobilization, leadership, and community development, and institution building.

4.1.3. Generic Challenges in Extension

The various roles that agricultural extension plays and the political, social, and administrative frameworks in which extension systems function all contribute to the general issues that the field faces. Regardless of the management system or expansion strategy, we think that concentrating on these common issues brings to light the topics that ought to be on the agenda for future developments in the field. There are common issues with extension functions; for example, many of the services provided by extension are toll or public goods, necessitate group effort, and frequently involve the government. We identify eight general issues.

4.1.3.1. Scale and Complexity

The reason these problems are at the top of the list is that they almost always reflect a "given" scenario. When considering the quantity, distribution, and variety of employees, farmers, and other clients and stakeholders, along with the mission and technique, it is possible to comprehend the scope and complexity of the extension work. We discussed the numerous clients of the few extension agents, many of whom are low-income and resource-poor. The management choices made by each of these millions of farmers will determine whether extension is successful or not. Many populations are difficult to reach and widely distributed (Chambers, 1983). Clients in underdeveloped nations typically reside far from information sources, have low levels of literacy, and have not received a formal education. Depending on their natural habitat, culture, farming/production system, and gender, they have different needs (Zijp, 1998). The various partners and stakeholders in the agricultural development endeavor, the variety of extension mandates, and the various sources of agricultural information and assistance for farmers are all examples of complexity. Extension decision-makers face increasing complexity because to the range of communication formats available, including mass media, computerized information technology, farmer-to-farmer extension, and individual farm visits. Coverage is the primary way that the magnitude problem manifests itself. Only 10% of prospective customers are being reached by many public services, and women make up a small portion of this group. To decide coverage and focus, magnitude interacts with appropriate operating resources, technology, and governance quality (Nagel, 1997). Establishing comparatively big, hierarchical, centralized public-sector structures with a large number of extension agents widely dispersed in rural areas, where their work is difficult to see or check, has been the most prevalent approach to the coverage problem. Large bureaucracies typically have a top-down management style that makes participatory, bottom-up approaches difficult to implement, and the hierarchy's multiple levels isolate decision-makers from the actual work being done in the field.

This intensifies the frequently hidden inclination of extension agents to give preference to more receptive consumers who are usually better off and more willing to take chances (Axinn, 1988). Biases of this kind have been noted in a variety of nations, including Tanzania, Estonia, and the Netherlands (Zijp, 1988). There is a natural conflict between being answerable to all farmer strata and establishing a rigid hierarchy that can oversee a system this size.

4.1.3.2. Dependence on the Wider Policy Environment and Other Agency Functions

Agricultural extension has an inherent difficulty in that, in order to accomplish agricultural growth, it must be integrated with other policy instruments, particularly in low-income countries (Van den Ban, 1986). Therefore, removing more obstacles to the agricultural sector's overall success—such as credit, technology stock, input supplies, price incentives, institutions, and human resource constraints—is essential to the effectiveness of extension investments (Purcell and Anderson, 1997).

A few examples of discriminatory institutional frameworks and agricultural policies include locking up arable land for low-productivity ranching, undervaluing, and wasting natural resources, underinvesting in the health and education of the rural population, discriminating against private sector initiatives in food marketing, and failing to maintain or invest in new rural infrastructure (World Bank, 1997).

Thus the influence of extension on productivity is limited by resource scarcity and ineffective functioning of complementing agricultural services. For extension organizations, coordination, and connections with supplementary agricultural services—particularly those with research, credit, input supply systems, and marketing organizations—are major issues (Axinn, 1988). The usefulness of expansion investment is frequently compromised by the inadequate consideration of this dependence.

4.1.3.3. Inability to Trace Cause and Effect

An additional common problem with extensions is the challenge of determining the causal relationship between their input and their effects. This challenge gives rise to a number of other innate issues, such as accountability, budgetary constraints, and political support. The relationship between extension activities and changes in 1) farmer awareness, knowledge, and adoption of certain technologies or practices; and 2) farm productivity and efficiency, profitability, input demand, and output supply is measured in order to assess the impact of extension. Numerous more elements that can cause confusion also have an impact on these same signs. Purcell and Anderson (1997) noted severe data constraints and an inability to include all the relevant elements impacting output outcomes in their analysis of World Bank agricultural research and extension initiatives. Farmers' decisions on how to use the resources at their disposal and their particular production response are influenced by a variety of factors. More in-depth econometric research is required to understand these links. Feder and Slade (1986) examined the challenges associated with monitoring and quantifying changes in farm husbandry expertise, agricultural productivity, and the operational effectiveness of extension agents as a result of more intensive systems that extension had established. They observed that it was impossible to compare the subject and control areas and that there was not enough time-series data to cover the post-project scenario. It is uncommon to find an optimal simulated experimental framework (before/after and with/without an intervention), as noted by Birkhaeuser et al. (1991). Depending on the level of analysis, second-best techniques have different kinds of biases. Studies conducted at the farm level are susceptible to issues with self-selection and the frequency of communication among farmers. There are estimate issues with aggregate effects of extension evaluated in a region related to confounding factors. The methodological difficulty is summed up by Axinn (1988), who points out that monitoring inputs is far simpler and less expensive than monitoring outcomes. The expense of gathering data increases with the sophistication and degree of extension indicators, and it becomes more challenging to establish a link between the chosen extension activity and changes in farm revenue or wellbeing. The fundamental issue, though, is that this is the exact data required to justify and increase the effectiveness of extensions.

4.1.3.4. Commitment and Political Support

In over half of World Bank-assisted free-standing extension projects, top government officials' lack of commitment has been identified as a factor negatively affecting execution and financing support (Purcell and Anderson, 1997). The government's inability to provide the required funding to operate extension systems is a major sign of this kind of unwillingness. According to Umali-Deininger (1996), 87% of programs classified as satisfactory and 100% of those rated unsatisfactory are affected by insufficient recurrent funding, which hinders the field operation of World Bank-supported extension initiatives. According to Roling (1986), terms of trade and bargaining power for the agriculture industry are linked to fiscal issues. Bias towards cities and a lack of knowledge about rural issues frequently contribute to a lack of dedication to agriculture and extension (Zijp, 1998). Because the attribution of output achievements to extension is so hard, another explanation for the lack of quick payback for politicians and policymakers for their support and commitment to extension is that there is none.

4.1.3.5. Accountability

According to Howell (1986), accountability has three components:

1. Extending performance in terms of its ability to deliver necessary services and pertinent information and its impact, effectiveness, or benefit/cost ratio.
2. Supervisors' accountability for field workers who are dispersed and often operate unsupervised.
3. Public sector staff accountability to fanners.

Specifically, the second and third characteristics are of relevance to us. Within the public sector, where there is a supply-driven, "top-down" extension hierarchy and agents report to ministry supervisors instead of fanners, the third accountability issue surfaced. However, agents become answerable to no one because even their supervisors find it impossible to oversee and assess their work due to the complexity of causality. Rent-seeking possibilities, on the other hand, give extension agents an incentive to concentrate on non-extension tasks that yield more readily observable outcomes, including credit and input invoicing. Numerous World Bank extension programs had signs of accountability issues, according to Purcell and Anderson (1997). The farming community's methodical involvement in problem definition, problem solving, and extension programs received little recognition. Technical support and farmer relations were severely hampered by staff attitudes and quality. In 48% of the programs that were rated excellent and 75% of the projects that were assessed unsatisfactory, the World Bank's 1994 ex-post evaluation of extension projects found "entrenched top-down attitudes." Accountability is a problem everywhere, not only in underdeveloped nations. According to Hercus (1991), the government of New Zealand received its agricultural extension services' reimbursement in terms of "activities not results, and, almost exclusively [in terms of] expenditure and hardly at all with outputs or efficiencies." The Agricultural Extension Service derived the mandate of extension on its own, and considered its charter to be the right to continue operating under the current terms and conditions in the absence of any objection or alternative definition from the representatives of the taxpayers."

4.1.3.6. Liability to Public Service Functions Beyond Agricultural Knowledge and Information Transfer

Since the extension service frequently serves as the most extensively dispersed grassroots government agent in the rural sector, there is always a desire to give it an increasing number of responsibilities. Public extension has historically been tasked with a variety of public duties, such as data collection, surveying, report writing, erosion management, and various regulatory duties like setting production quotas or using pesticides. Feder and Slade (1993) point out that as governments became more involved in the rural economy, the agricultural field service was extended to include a number of new duties in numerous nations. The government's broader efforts to alter rural mindsets and encourage community self-reliance include the distribution of agricultural expertise through rural extension (Oakley, 1997). The government views the village-level worker as a flexible, reasonably priced administrative tool that may provide multifunctional services to underprivileged communities and assist them acquire greater support. We identify both "pull" and "push" elements contributing to this invasion of extension services. The "push" refers to the incentive for other organizations to employ extension since it is the most extensively used tool for interacting with rural populations. The "pull" is that agents are eager to take on additional responsibilities, particularly distributing inputs, as these activities frequently give them more clout with farmers and offer chances to collect rent, which helps to offset their meagre pay. Field personnel can accomplish this since they are unable to link the source and effect of the extension. It is generally discouraged to assign input supply chores to extension workers since time spent on non-educational tasks can limit the potential educational impact by as much as 25% (Swanson et al., 1990). But as Van den Ban (1986) points out, if demand is low or "large" farmers control farmer groups, there may be issues with using private sector or farmer organizations as a substitute to extension delivery of these services. Extension needs to coordinate even in the case of distinct agency organization.

4.1.3.7. Operating Resources and Fiscal Sustainability:

Insufficient public funding to effectively operate services is a typical occurrence in World Bank-assisted freestanding extension initiatives, according to Purcell and Anderson (1997), with 76% of projects having an uncertain or improbable sustainability assessment. Training and visit (T&V) extension systems, which usually cost 25–40% more than multipurpose extension (Feder and Slade, 1993), actually make operational budget difficulties worse by expanding manpower, according to Antholt (1994). The fundamental challenge of cost recovery is the general issue with extensions. A large amount of information that is distributed through extension is considered a "public good," and as it is difficult for people to pay for distribution expenses, direct governmental support is required. In addition to uncertainty concerning the "model issue" (Rivera, 1991) and the state's function (is it controller, funder, organizer, or implementer), there is a lack of political commitment and support. A huge extension service serving a large number of farmers with a large staff that is expensive to run is partially to blame for the issue. Salaries have high fixed costs because running expenditures are viewed as a leftover and are therefore susceptible to cutbacks in the budget. If a government must reduce its operating budget, there will undoubtedly be an impact on efficacy as extension agents will be unable to reach the field, which will lead to issues with morale and operations (Axinn, 1988). According to Howell (1985), extension organizations followed a cyclical pattern in which they reduced operating expenditures when budgets were tightened but increased staffing when funding was available. According to Ameur (1994), the issue is also seen as a vicious cycle involving decreased staff motivation, training, and competence, ineffective operations, shortened services, and worse outcomes.

4.1.3.8. Interaction with Knowledge Generation

Many would contend that the message's quality is the "bottom line" of extension. However, one of the most prevalent limitations in extension and a significant one in rainfed, resource-poor regions is the lack of innovative or relevant technologies required to increase production (Axinn, 1988; Anderson and Purcell, 1997). Inadequate research-extension ties negatively impacted a significant number of the programs examined, according to the World Bank's 1994 ex-post study of extension initiatives, and a lack of appropriate technology was a concern that was even more prevalent. The fundamental issue is that extension distributes data and recommendations produced by a knowledge-generating system that is typically not managed by extension. This results in an essential dependency, which is not always problematic in and of itself. Nonetheless, research, and extension frequently have a tendency to vie with one another for funding and influence, failing to recognize their place in a larger agricultural technology system. Additionally, research, and extension have a propensity to explore within national borders for solutions, even though pertinent data and technology may be easily accessible beyond such fictitious borders. According to Kaimovitz (1991), the following barriers exist for successful research-extension connections:

1. **Historical Perceptions:** Researchers and extension continue to be overlooked by policymakers despite their strong interdependence. Because of its past, Extension's credibility as a research partner is still tarnished. Extension agents dislike the condescending behavior that comes with scholars' generally superior standing. When it comes to allocating time, resources, and managerial attention, linking activities are ranked lower than core activities by both research and extension.
2. **Coordination:** Resistance to coordination is perceived as limiting autonomy by both sides.
3. **Goals May Differ:** It's possible that the two groups have different objectives. Although the formal purpose may be to supply farmers with agricultural technology, it's possible that researchers are more focused on writing scientific papers than on doing research that is useful to farmers. If extension agents share the credit and input, they might be rewarded more.
4. **No Pressure:** Cooperation may be hampered by the absence of well-organized outside pressure groups, such as donors, farmer organizations, national legislators, or private businesses.

4.1.4. Generic Challenges Interactions

It is impossible to view the common issues we have found in isolation from one another. They frequently reflect competing imperatives and are connected to one another. There may be pressure to emphasize firm management control while guaranteeing bottom-up participation, to maximize coverage while minimizing expenses, and to improve human resource competency while taking use of the mass media's potential for cost savings (World Bank, 1990). We have highlighted significant causal links, including the impact of complexity and scale on the size and hierarchical structure of national extension organizations, as well as the implications for political support and the impossibility of establishing a cause-and-effect relationship. Despite the fact that issues with staff quality and low morale, inadequately trained and motivated individuals, and a dearth of qualified, competent, and trustworthy staff have all been documented (Antholt, 1994; Axinn, 1988; Nagel, 1997), we see these issues as indicative of the more general issues that have been recognized. The interactions may also be connected to Ameur's (1994) vicious cycle of financial difficulties, reduced services, and inefficient operation. According to Kaimovitz (1991), low pay, few operating resources, and a murky mandate contributed to a general downturn in morale as well as challenging supervision, poorer extension work, greater employee turnover, and diminished credibility. Extension finds it more challenging to acquire resources as a result of these issues. Researchers who saw extension as ineffectual and hence reluctantly cooperated were directly impacted by this crisis in extension. Other noteworthy exchanges consist of:

1. The potential irrelevance of technology generation and its "top-down" orientation. If biassed, it can also skew coverage and concentration, ignoring more marginalized, resource-poor places as well as weaker target populations and their crops (Axinn, 1988).
2. Unable to determine coverage, cause, and effect. Particularly for resource-poor farmers, there is little incentive to work harder to reach more if extra effort is not acknowledged or rewarded. Instead, field employees might concentrate on larger farms and high-input technologies.
3. Sustainability and accountability. An extension agency can only get a limited understanding of how willing consumers would be to assist without farmer participation (Ameur, 1994).
4. Insufficient funding and financial stability (as the independent variable), as well as staff morale, coverage, relevance, and responsiveness, which in turn affects accountability, and knowledge creation.

4.1.4.1. Overcoming Generic Problems

In response to the general issues mentioned in the preceding section, policymakers have developed a variety of institutional innovations. But the focus of these inventions has shifted over time. In the past, communication was prioritized over farming systems and constraints. Subsequently, the focus turned to enhancing the administration and structure of the current public extension services. In the last few years, efforts have been made to find a solution for the budgetary sustainability issue.

The idea that not all collective action needs to be planned or carried out by government institutions is a concept that is used in many developments. Consequently, community-based collective action offers numerous benefits that transcend general issues. Moreover, not all facets of extension's function are pure public goods; as a result, some developments concentrate on dividing components that can be privatized and subjected to market discipline, so resolving the motivation issues associated with the provision of public services.

4.1.4.2. Innovations and Modifications

We have identified eight generic difficulties, and our analysis focusses on eight inventions and attempted adjustments to address them:

4.1.4.2.1. Improving Extension Management

The most recent FAO reference manual on extension (FAO, 1997) acknowledges the significance of enhancing extension management. Since the training and visit (T&V) system of extension is a well-documented, extensively utilized field model of an intended systematic, comprehensive improvement in the organization and management of public extension, it is cited as an example of improving management. There are many more ways to formulate management concepts; but, their use has been less systematic and recorded less thoroughly. T & V has been explored or implemented in national systems in at least 76 nations across all major global regions since the late 1960s (Umali and Schwartz 1994). Purcell and Anderson (1997) assessed 90% of World Bank agricultural extension projects that used this system. According to Kaimovitz (1991), T & V was implemented in reaction to a historical crisis that public extension experienced in the 1960s and 1970s over efficacy, credibility, and morale—all of which were caused by generic issues. The designers of T & V emphasized that in order for it to be effective, a few essential elements had to be maintained: professionalism, a single line of command, effort concentration, time-bound labor, field, and farmer orientation, frequent, and ongoing training, and close ties to research (Benor et al, 1984). Many modifications have been made over time, but the original design—which we consider to be an institutional innovation that was induced—is being discussed here. The design called for village extension workers (YEWs) to advise farmers on prices, availability of necessary inputs, and market conditions. It also called for YEWs to report to supervisors on actual availability and farmer responses, without eliminating reliance on outside factors and other agency functions. This would have made it possible to modify the instructions for the extension. The stringent schedule of contact farmer group visits and the tight, line-of-command oversight structure were intended to provide accountability. Staff members who dedicated their entire attention to information distribution tasks were freed from pressure to do non-agricultural or non-extension-related work. Lastly, T&V designers structured research-extension ties incorporating ongoing input on farmers' problems and regular training in an effort to address the issue of contact with technology generation. In actuality, T & V exacerbated a few common issues and were unable to resolve others. Because contact farmers were not representative of the farming community, the contact farmer coverage strategy often encountered difficulties (Nagel, 1997). Strict scheduling made it possible to monitor and check in more closely with field workers, but it did not solve the issue of connecting input to impact. It was not possible to completely eliminate the reliance on other elements related to rural development, and the extension system's independent status under T & V initiatives did not assist in resolving issues with program coordination. Despite certain achievements, T & V has not resulted in consistent political commitment or support in a significant number of countries. Axinn (1988) and Nagel (1997) noted that there was still a tendency to overlook participation features and a lack of established accountability to farmers. The most notable drawback of T&V as a national system has been its "highly questionable sustainability" (Antholt, 1994) because of the significantly higher staffing levels and ongoing and operating expenses (Axinn, 1988).

4.1.4.2.2. Decentralization

Because diversity becomes more tangible and diverse ways to extension can be explored as the local level becomes accessible, decentralization has been referred to as "the first step on the long road to privatization" (Ameur, 1994). Administrative, political, and budgetary devolution of program and financing decisions, as well as personnel accountability to local units, are all aspects of decentralization. The degree of localized political and social democratization determines its effect. In particular, the following are the main variables influencing how effective decentralization is:

1. The existence of an elected, representative local government and a central government willing to actually decentralize,
2. The ability of local government to raise revenues, and
3. MOA capacity for efficient quality control and monitoring (Garfield et al., 1996).

If these conditions are satisfied, a variety of measures, including developing local capacity for farmer participation in extension programming, housing extension agents locally and holding them accountable to farmers' associations, and creating funding and resource mobilization mechanisms, may be implemented.

The top-down structure and functioning of a public service bureaucracy can be changed by decentralization, which can also have a positive impact on a number of the general issues associated with extension if it is executed successfully and local government receives adequate funding. The number and degree of "local" nature of local government entities that perform extension functions inside a nation determines the scale and complexity of the problem. The possibility of improved connection with other local projects lessens Extension's reliance on other agency activities. Democracy in local government improves responsiveness (a component of accountability), political support, and commitment. This gets around the difficulty of establishing a cause and effect relationship to some extent since the locally elected administration has an interest in ensuring client happiness. However, there is more room for local politics to meddle in areas of technicality. Building on significant national decentralization initiatives that are typical in the region, decentralization experiments in public extension services have been conducted in a number of Latin American nations (World Bank, 1997). The Colombian government decentralized extension to all but 10 of the 1050 communities in the nation between 1989 and 1993. Each was obliged to establish its own Unidad Municipal de Asistencia Tecnica Agropecuaria (UMATA), an extension agency that offers small farmers technical support on a variety of topics. In addition to employing 3500 technical and professional staff, the 1040 UMATAs offer free extension services to 450,000 small farmers, or 28% of all small farmers in Colombia. Up until now, practically all expenses have been covered by the federal government through a convoluted system of matching grants and budget transfers to municipalities.

According to increased staff inputs, Colombia's decentralization has so far enhanced coverage and maybe responsiveness. In conjunction with staff increases, the number of towns covered has increased by over 300%, beneficiaries have increased by over 250%, the beneficiary/technical staff ratio has decreased by 17% to 129, and the cost per farmer has decreased by 10%. Small farmers claim that employees of municipalities are easier to reach than those of the Colombian Institute for Agriculture and Livestock (ICA), which was formerly overseen by the MOA. But at least thus far, the decentralization has complicated program and staff quality control (accountability and monitoring), intensified local political meddling, severed research-extension ties, and raised staffing and overall costs approximately proportionate to the expansion in coverage.

The total cost has climbed by more than double due to a factor of 3.5 in personnel numbers. With over 1000 decentralized units, ensuring service quality has become more complicated, according to Garfield et al. (1996). Another criticism of the decentralized system is technology generation. Recently, a pilot project in Venezuela funded by the World Bank for agricultural extension started with decentralization innovations that could mitigate some of Colombia's possible issues. One of the pilot projects involved hiring a university or NOO on a subcontract basis to set up an extension team in each of the five trial municipalities. The main project contracts private extension consultants, NGOs, and universities to manage the extension service provision; it also provides for cost-sharing between the national, state, and municipal levels of government as well as the beneficiaries themselves. The main project decentralizes planning and implementation of extension to the municipality level and creates farmers' associations to oversee the municipal extension service. Farmers' contentment with the assistance they currently receive from extension agents has been substantiated by field visits.

4.1.4.2.3. Single Commodity-Focused Extension

The public, parastatal, commercial, and social sectors—agro-processing and marketing companies, farmers' associations, and others—have all used commodity-specific extension (NageI, 1997; Umali and Schwartz, 1994). Focus is frequently placed on a single commercial or export crop connected to well-established channels for selling or processing, or on a single facet of farming, like dairying or livestock (Purcell and Anderson, 1997; Axinn, 1988). Commodity-specific extension is unique in that the majority of the production and marketing system's components—research, credit, product marketing, input supply, extension, and occasionally price assurance—are vertically integrated. By internalizing the complementary services, it thereby addresses the general issue of reliance on broader policy and other agency tasks in the most efficient way possible. Concentrating on a single commodity narrows down complexity and scale. The issue of linking cause and effect is largely resolved by this organizational model, which is a very small organization that incorporates every facet of the commodity process. For similar reasons, consistent pay, benefits, and working conditions for employees easily ensure staff responsibility. Liability to other public service duties is decreased by concentrating on specific commodity problems. The small-and-focused strategy is comparatively cost-effective, and fiscal sustainability is attained by imposing taxes on product sales or by including cost-recovery into the price of products or inputs. Lastly, the vertically integrated structure virtually guarantees a "technology fit." Extension services are offered by agro-processing and marketing companies to their farmer suppliers in an effort to lower the risks associated with input supply, lower post-harvest losses, and enhance output consistency, quantity, quality, and timeliness. in the marketing and agro-processing. Nowadays, extension services are frequently a crucial part of high-value commodity production contract growing schemes. Umali and Schwartz (1994) offer a plethora of global instances of the diverse array of goods advertised in this fashion. They also provide numerous instances of cooperative commodity businesses and farmers' associations. The commodities strategy is not without its drawbacks. The benefits are mostly determined from the standpoint of the commodity organization, as noted by Nagel (1977). The technique obviously does not work as well in cases where farming is not a monoculture, as the narrow concentration on commodities leaves many unmet demands for the remainder of the agriculture sector and the extension system. Consequently, concerns about the commodity's relationship with other elements in the smallholder production system, scale, and coverage, and other public service roles are disregarded (Purcell and Anderson, 1997). Unless the commodities organization is under the supervision of a farmers' association, accountability to farmers is frequently in doubt (Axinn, 1988).

4.1.4.2.4. Fee-for-Service Public Provision

Certain public extension services have started charging for their services, mostly as a means of covering their costs. A portion of the fees are paid by farmers, but the government also contributes money under contract. Commercialization can occasionally be seen as a step on the way to privatization. The main general issues this method addresses are those of accountability and fiscal sustainability. The main result of introducing fees for services, along with performance-based staff commissions, is to improve accountability and efficiency by fostering a more professional and client-oriented relationship between extension agents and farmers. This is in addition to recovering costs from farmer clients who can afford to pay for information and advice. The example from New Zealand that follows shows that integrating extension with other activities and establishing cause and effect can have some positive effects. The introduction of user fees will only have a positive effect on the scale (coverage) problem if it is combined with client market stratification and the provision of unique services for farmers who are not as commercial. Without this safeguard, fee-for-service expansion would probably make the general issue of coverage in low-income nations worse. Additionally, as is the case with other privatization-related changes, it would raise issues with political support and commitment. New Zealand's and the UK's advisory services were examples of public extension providers that increased the percentage of their services that were chargeable. In 1986, New Zealand started its cost recovery initiatives. In 1995, the government sold Agriculture New Zealand (AgNZ), formerly known as the Ministry Advisory Service, to Wrightson's, a farm service and stock trading company (Mavromatis, pers. comm.). Half of the funding comes from government contracts and training; the other half comes from extension agreements with individual farmers, farmer organizations, research institutions, and agribusiness firms. In 1987, ADAS, the UK advising service, started charging for services. In 1992, the agency gained greater autonomy under government ownership; in 1996, it stopped receiving government subsidies; and in April 1997, management, and employees bought the company to enter the private sector (Griffis, 1996 and pers. comm.). Cost recovery was the goal when fees were first implemented in 1987, and farmers were expected to pay for advice that directly improved their companies. Although it wasn't a declared goal at the time, privatization did become one in 1994–1995 as it became evident that full cost recovery was achievable (Griffis, pers. comm.). The Ministry of Agriculture, Fisheries, and Food (MAFF) of the United Kingdom government continues to be the biggest client of ADAS. It handles a variety of tasks related to policy creation, execution, and oversight, including contractual extension. The clientele of the commercialized ADAS shifted as it sought out new business opportunities, concentrating on paying farmers and corporate clients in the agricultural sector. Due to commercial demands, there was less time available for extensions, and they were quickly restricted to the tasks that MAFF needed completed and would pay for. Consequently, there was a modest decline in the total number of farmers reached. In the UK, farmers had access to a wealth of information from other sources, thus this was not seen as a major issue.

The increased activity of consultants has enhanced their performance and job satisfaction, even though no formal assessment of the impact of the New Zealand modifications has been conducted (Mavromatis, pers. comm.). The expansion in the number of farm consultants across the nation could be another indication that commercialization has improved coverage overall. Mavromatis says that a variety of organizations financed mostly by producer levies have been established to fill the void in public good extension left by AgNZ's privatization. Citing Hercus (1991), Rivera, and Cary (1997) point out that by integrating extension workers across the whole production, processing, transporting, and marketing chain, commercialization in New Zealand enhanced accountability and reduced the economic load on the public sector. It also moved away from just participating in activities and towards a greater focus on the client and identifying and achieving results. Griffis outlines improved attention, more professional connections with clients, more job satisfaction for the majority of employees, and decreased government costs for ADAS. The relationship between information development and coverage for small farmers in New Zealand continued to raise some unfavorable issues. The efficiency and emphasis of research agencies and advisory consulting work improved. AgNZ had contracted with commodities research organizations and the national Foundation for Research, Science, and Technology to carry out certain "public good" technology transfer programs. However, communication between the various organizations deteriorated, and farmers' input to science providers decreased. When AgNZ was a government-funded extension agency, less wealthy farmers, in the perspective of general manager Mavromatis, tended not to attend field days; however, farmers are exposed to a wider range of viewpoints these days. Reduced services have also been provided to low-income farmers in the UK because some of them have not been persuaded that the advantages of using ADAS services exceed the costs, especially in light of the availability of information from other sources in the country. Howell (1986) and Rivera and

Cary (1997) point out that user fees have a limited capacity to support public extension services in developing coteries. Reorienting and retraining extension workers, creating cost-accounting methods, and collecting user fees may all present obvious challenges (Griffis, pers. comm.). User fees for "common good" general extension information would be difficult to implement and would even lower general extension participation in economies that rely primarily on subsistence farming (Howell, 1986). According to Umali-Deininger (1996), market-oriented farming operations—especially those in areas where medium- to large-scale farmers predominate—will almost entirely drive demand for fee-based extension services. As a result, she suggested that implementing partial cost recovery could help to create a more demand-driven system and act as a crucial stage in the process of creating a market for the extension of fee-for-service contracts. Projects in Chile, Mexico, Venezuela, and Nicaragua that are supported by the World Bank feature cost recovery and user co-financing elements. According to Wilson (1991), one way to reduce costs in Mexico is to divide up the client market based on income levels and either gradually advance higher-income producers to private extension services or demand a larger level of cost-sharing. By reducing general fiscal and liability issues, strategizing for cost recovery frees up public funds for a "extension safety net" that focusses on low- to middle-income producers in priority areas (Umali-Deininger, 1996).

4.1.4.2.5. Institutional Pluralism – Mobilizing Other Players

This category of innovations aims to develop a more diverse system of supplementary extension services that can respond to and reach a range of farmers and farming systems (World Bank, 1997). Some of these agreements, like subcontracting and cofinancing, have a tight line separating them from fee-for-service extensions. With a few exceptions, the focus of the fee-for-service extension was mostly on cofinancing public provision; in this case, however, the focus is on shifting more towards private provision. We give examples of subcontracting agreements with private sector enterprises, including voucher and coupon systems, cofinancing, and cooperative arrangements with NGOs and farmer organizations, after analyzing the general ways in which innovations in this category overcome some generic challenges. Pluralistic arrangements have the potential to assist in resolving two basic generic issues: tying cause and effect together and providing accountability or an incentive to provide high-quality service. These issues can be resolved by incorporating a range of stakeholders in the process of forming contracts and cooperative partnerships. In subcontracting arrangements, the contracting process strengthens the provider's client orientation, and the farmer's power grows as a fee-paying customer. These agreements typically incorporate multiple stakeholders in the development of the contract terms of reference, competitive bidding, and direct farmer involvement into the indicator design. "Getting around the institutional inefficiencies associated with public delivery" may be the driving force behind subcontracting (Umali Deininger, 1997). These often comprise a range of temporary issues that have an impact on employees' performance but are less common in the private sector. Involving nonprofit organizations may enhance coverage fairness, cost-effectiveness, and responsiveness even further. The challenge of institutional pluralism lies in the central government's need to adapt to having less direct control over staffing and programs. Additionally, as new technologies are built, the complexity of financial and administrative management may initially grow. It might take more time and resources to keep an eye on the quality of the services. The issues of coverage, accountability, fiscal sustainability, and interaction with knowledge generation can all be resolved by mobilizing additional players through these approaches, provided that these challenges can be met and complementarity between public and private sector roles can be achieved. There are several guiding concepts for advances in this area. Separating public finance from public delivery comes first (Zijp, 1998). Second, governance must alter as a result of plurality. There are new organizations participating, as well as new institutional arrangements including public-private partnerships. Encouraging all stakeholders to voice their opinions and interests and take appropriate part in the planning, execution, and assessment of extensions is a fundamental governance principle.

Third, pluralism acknowledges that the government should make broader investments in the agricultural knowledge and information system (AKIS) as a whole rather than focusing only on public sector extension services in order to address the wide range of demands and circumstances in the farming sector. Each of the aforementioned principles implies a significant shift in the role of ministries of agriculture away from service delivery and towards the creation of an environment that supports policy, the coordination and facilitation of other actors' work, and support for farmers in contract negotiations, quality control monitoring, and financial control. Several Latin American nations are attempting to address issues like fiscal sustainability and poor client orientation by integrating the private sector into extension systems, realizing that complete privatization of agricultural extension services is frequently not feasible (Umali Deininger, 1996). One option is to contract for just certain services, such as video creation and staff training, as is the case in Morocco. There have been attempts at more extensive subcontracting of extension services in a variety of nations, including Estonia, Turkey, Madagascar, Costa Rica, and Mexico. Colombia has employed coupons tied to bank loans for agriculture, allocating a specific portion of the loan for extension services (Rivera and Cary, 1997). A novel approach to subcontracting, extension vouchers were introduced in Costa

Rica and Nicaragua. In Nicaragua, farmers who received vouchers may select between governmental and commercial extension providers (Umali-Deininger, 1996). In addition to providing funding, the government continues to have certain regulatory authority over extension providers. The sorts of farmers and technological levels in Costa Rica that require high- or low-intensity technical support determine the vouchers. Beneficiaries are expected to continue using only private technical support when the initiative is completed. Which specific farmers should be graduated from the program is a recommendation that the private extensionist will make to the MOA. To determine whether these coupon and voucher systems come at a substantial expense, there is yet not enough information available. The history of voucher programs in the sphere of education (West, 1996) indicates that significant administrative expenses could jeopardize the long-term viability of the voucher program. In Ecuador and Honduras, cofinancing seeks to supplant state extension services with a "technology transfer market" that is co-financed by the beneficiaries and the government, where private providers would vie for market share. In these situations, farmers are categorized into three groups: small-, medium-, and large-scale producers. Different programs and graduated copayment scales are created for each of these farmer categories. Cooperative agreements with universities, commodity boards, and commodity foundations are examples of collaborative partnerships with the NGO and nonprofit sector (Umali and Schwartz, 1994; Nagel, 1997). In underdeveloped nations, nongovernmental organizations are a very common partner in agricultural extension, usually concentrating on areas that the government does not effectively serve. Numerous non-governmental organizations (NGOs) aim to be democratic, responsive, participative, affordable, community-based, and targeted at difficult-to-reach target groups. NGOs do not, however, consistently adhere to their tenets. Certain non-governmental organizations prioritize their own objectives and answer more to outside sponsors than to the people they are supposed to assist (Farrington, 1997). As a result, caution must be taken to make sure that agreements for partnerships with NGOs build on their advantages and minimize their disadvantages.

4.1.4.2.6. Empowerment and Participatory Approaches

One of the most significant steps towards holding extension companies accountable to their clients for performance—and addressing a major general weakness of these organizations—may be the beneficiaries' evolving control (Antholt, 1994). By fixing this problem, the flaw resulting from the inability to determine cause and effect is removed. As beneficiary organizations mandated by some decentralized, cost-recovery, subcontracting, and cofinancing arrangements, we have already mentioned farmers' groups. One important component of NGOs' client-based extension initiatives is farmer organizing (Nagel, 1997). Additionally, as we noted in the previous subsection, farmers' associations that are structured along commodities lines really offer their members extension services (Umali-Deininger, 1996). Certain farmers' organizations that are founded on commodities have achieved great success; one such example is the Indian dairy business (Chamala and Shingi, 1997). A plethora of knowledge and insights on involvement and farmer control attest to possible benefits on general issues. Their influence, however, varies according to where a given initiative falls on the "participatory continuum." In their study of the subject of rapid/participatory rural evaluation, Pretty, and Volouhe (1997) listed over 30 different names and terminology for various participatory learning and action systems that have emerged in the last 10 or so years. "Self-mobilizing" techniques are used in many of them, including sampling, interviewing, and dialogue, group, and team dynamics, visualization, and diagramming. The examples they provide make it abundantly evident that the deliberate application of these techniques, which place a strong emphasis on self-evaluation and monitoring, significantly improves the degree of trust, understanding, and connections between the various agencies and actors involved in a rural situation, with the farmers at the center of the process. In their evaluation of the lessons learnt in starting and growing farmer organizations, Chamala, and Shingi (1997) affirm the legitimacy of extension roles in community organizing, issue solving, education, and farmer empowerment. The "participatory approach" has benefits and drawbacks, as noted by Axinn's analysis (1988).

Participatory approaches yield positive outcomes for the majority of general extension concerns. Regarding the issue of size and coverage, participatory models generate farmer leaders—including women—with suitable local backgrounds who can efficiently carry out a variety of extension agent responsibilities (Axinn, 1988; Russell, 1986). According to research, farmer dependence on outside inputs is decreased and complementary services are more closely tailored to farmer requirements when participatory approaches are used (Axinn 1988; Roling and Pretty, 1997). In Europe, the establishment of farmers' unions increased governmental support and enhanced the integration of supplementary services (Roling, 1986). Through farmer-led experimentation, analysis, and feedback, participatory techniques also benefit the process of tracking cause and effect (Axinn, 1988). Accountability is one of the main benefits of involvement. Raising farmer awareness and confidence is how "automatic quality control" is accomplished, according to Axinn. Additionally stressed are the qualities of ownership (Chamala and Shingi, 1997) and established trust (Pretty and Volouhe, 1997). Local resource mobilization improves fiscal sustainability. Efficiency and cost-effectiveness are attained by employing pertinent techniques that prioritize farmer requirements that are articulated and local people taking on a large number of extension responsibilities (Axinn, 1988). By integrating indigenous knowledge with feedback into the agricultural knowledge system and identifying and validating relevant technology and technology generation needs for farming systems, participation improves the interaction with knowledge generation (Axinn, 1988; Chamala and Shingi, 1997; Purcell and Anderson, 1997). Examples of the significant roles that farmers' organizations can play in arranging and financing agricultural extension include France, Norway, and Taiwan (*see* Ameur, 1994; Umali and Schwartz, 1994; Nagel, 1997). 90% of Taiwan's farmers are members of farmers' groups, which are arranged at the township, county, and provincial levels (Nagel, 1997). The government has established an extension program that encompasses robust and formalized connections with research and additional services. At the township level, extension is managed by agents hired by the farmers' groups and is primarily funded by the farmers themselves. The Agricultural Experimentation Regional Consortia (CREA) farmer group movement originated in Argentina and has now expanded to Uruguay, Brazil, Chile, and Paraguay. It is a prime example of a farmer-organized, completely self-sufficient extension (Tobar, 1996). Not every attempt at participation results in an entirely positive experience. Effective local engagement frequently requires decentralization as a virtual requirement. Conversely, in traditional agriculture where participation is largely passive, there's a chance that the wealthy will pocket the benefits that are provided (Howell, 1986).

4.1.5. Privatization

The word "privatization" is frequently used in a misleading way. In its purest form, privatization means that all ownership is transferred from the government to a private organization (often through a sale), with the private organization bearing all expenses and earning all profits (Rivera and Cary, 1997). Most of the time, governments haven't truly privatized this aspect of agricultural extension services. The private sector is incentivized to furnish "better-off" commercial farmers and members of private associations, for whom extension service delivery is profitable, with private and toll-good information. It makes sense to engage the private sector to supply investment capital and services in regions where commercial farming and farmers producing goods with a marketable product predominate (World Bank, 1997). Strong incentives exist for input providers to offer guidance on a variety of livestock and crop-related tasks. This kind of technical information may become even more important given the growing commercialization of agriculture in many emerging nations.

In nations with a sizable population of small-scale subsistence farmers, fully privatized extension is not economically possible (Umali-Deininger, 1996). Under such conditions, public sector funding is still crucial, combined with different forms of cost-recovery, cofinancing, and other institutional structures that are transitional and suitable for the rate at which agricultural structure and commerce are changing. There will be good effects on certain general problems if privatization is appropriate. In order to overcome the cause-and-effect issue, all privatization initiatives record increases in accountability, which are often measured in terms of client orientation and satisfaction. Initiatives in the private sector are obviously unaffected by other public service obligations. By meeting the demands of farmer clients who can afford to pay for the information, all privatized efforts promise better efficiency, cost-effectiveness, and decreased public sector costs, hence resolving the sustainability and dependence on budgetary allocations issues. To have a product that can be sold, there are incentives for private extension providers to stay in close contact with knowledge generation organizations. Privatizing extension has a mixed effect on other general issues. If private extension is used excessively, lower-value crops and less commercial producers may be neglected. To mitigate this danger, stratification, and distinct, publicly-funded focused initiatives are required. Furthermore, the complexity of delivering a service that is ideal for society and the environment is not addressed by privatization. The majority of analysts advise open communication between all parties involved and a careful, progressive approach to privatization within the framework of a well-defined goal and plan. More commercial farmers, for whom technology packages already exist and extension is essentially a delivery function, may be the target market for privatization to begin with. It can start small and grow over time, taking farmers to the point where private sector services can meet their needs for extension in the future or they can pay for it, freeing up public resources to serve new cropping systems and clientele, including more marginalized groups (World Bank, 1990). There are numerous such examples where extension services have proven to be beneficial for private organizations. In the 1970s, Chile fully privatized its extension system, compelling commercial farmers to use privately held consulting firms for extension services (Umali-Deininger, 1996). Since then, the country has been working to streamline services for various farmer groups. The move "to full private sector delivery" did not have a significant impact on major commercial farmers, but it excluded small-scale and subsistence farmers from the extension market. The Chilean government was forced to aggressively direct extension assistance towards organizations that produced less that could be sold as a result. With help from the World Bank, two target groups were identified and given particular programming. Through a unique approach, a subgroup of farmers who were thought to have the resources to eventually achieve self-sufficiency and market surplus output were given access to a producer-oriented extension package. This recipient group contributed 15% of the possible 50% of the extension expenditures. A more farm family focused, "basic" service was provided free of charge to the second subgroup of poorer subsistence farmers, with an eventual payment of 15% of overall expenditures suggested. Private consulting businesses under contract provided services for both initiatives. Wilson (1991) points out that the objective is to enhance the contributions from commercial and family farms. The main idea of cost reduction in Chile is to focus on providing free services, restrict the duration of coverage to all recipients, and collaborate with groups to minimize expenses. A company must meet technical and professional personnel requirements, submit a bid for a contract, and consent to have its operations monitored and assessed by a designated governmental body in order to be eligible for the Chilean program.

According to Ameur (1994), these programs reach a significant number of small-scale farmers; the farmer-to-agent ratio is typically as low as 48 to 1; the consulting firms only work on technology transfer; and the farmers enter into agreements with their consulting firms to pay up to 30% of the program's cost. The government organization drafts the contracts' terms of reference. The National Agriculture Society in Chile is currently entirely privately supported and runs a recent program for medium- and large-scale farmers. Halfway through a two-year gradual privatization process, the Dutch government established a foundation with equal representation from farmers' organizations and the government on the board of its public extension agency, DLV (Rivera and Cary, 1997).

A deal was made with the Dutch government in December 1997 to end all direct bilateral funding contracts with DLV in three more years. Government contracts pertaining to extension projects will be tendered in an open market, with competition among all interested companies, starting in 2000. After that, the foundation will become a stock company, with the MOA temporarily holding shares in addition to employees. The Dutch privatization was revolutionary in certain ways.

According to Tacken (1996), DLV had to terminate or retire early from 50% of its initial workforce. This was a highly contentious process as DLV overhauled its organizational culture. Tacken lists a variety of advantages of the privatization, such as a 50% decrease in overhead. In just three years, client satisfaction increased by 40%. This increase was ascribed to increased specialization of advice, quality control, and enhanced client orientation. Prior to their privatization, extension programs had little effect, were strongly tied to the objectives of national policy, frequently ran counter to the interests of farmers, created contradictory roles, low staff morale, high employee turnover, and a lack of trust among farmers. The service overused mass media in its delivery and was neither needs-driven nor problem-oriented. Following privatization, farmers' concerns gained prominence since they were well-represented on the sectoral councils and DLV board. As DLV kept track of each employee's performance, attitudes among the staff regarding productivity, efficacy, and job satisfaction increased. Another recent example of a national attempt at privatization is Albania. In order to supply 600,000 recently founded small-scale private farmers with inputs and technology, the government established a competitive statewide network of 300 dealers with support from the World Bank (Schultz et al., 1996). The program specifically addressed the general issue of reliance on broader policies and other agency responsibilities since the government executed policies in a coordinated and uniform way to facilitate the growth of the private sector.

Furthermore, it financed the Albanian Farmers Union and a private sector trade association called AFADA, which both pushed the government to lower import taxes and do away with the value-added tax on fertilizer. Reliability of agricultural data, a weekly market price database, and other regulations, services, and fair norms for competition in the fledgling private sector were the exclusive focus of the restructured MOA position.

Accountability was another general issue that privatization helped to resolve through the enduring client ties private dealer businesses build. A direct approach to the private sector was needed to address the general issue of political commitment.

This method avoided vested power and rent-seeking interests, which slowed down and interfered with the reform process. The overarching issue of interaction with knowledge formation persisted. According to Schultz et al. (1996), while privatization made it easier for technology to be transferred and adopted, less government funding also led to competition rather than cooperation within the knowledge system, which made it more difficult to communicate with suppliers, researchers, educators, farmer organizations, and private consultants. Lastly, as we've already discussed, farmers' associations offer their own private extension services to members, frequently along commodity lines.

The involvement of private consulting firms offering extension services in Argentina, Brazil, Colombia, Mexico, Uruguay, Korea, and Taiwan is further documented by Umali-Deininger (1996). Approximately 2000 private consulting businesses served the specialized needs of the commercial livestock industry in Brazil in 1988. Consultancy firms in Asia typically focus on plantation crops and are staffed by former plantation managers and experts.

4.1.6. Interconnecting Rural People and Using Appropriate Media

The emergence of the information age has inevitably raised interest in the potential for cutting-edge information technologies (IT) to be applied creatively to improve the delivery of extension services. The role of more conventional extension techniques like mass media, group discussions, field trips, demonstrations, and exchange visits is also taken into account when choosing acceptable media for extension (Campbell and Barker, 1997). The most direct link between innovations in this area and solving the general issues of scale, complexity, and fiscal sustainability is the cost-efficiencies connected to specific mass media. The aforementioned privatization initiative in Albania made considerable use of radio, television, and a widely distributed monthly agricultural bulletin as means of disseminating information, raising public awareness, and influencing public opinion. The general issue of relying too much on broader policy was mitigated by the media campaign in favour of the revised MOA function.

According to a study by Wete (1991), radio, and print advertising, which are the most conventional methods, are the most suitable and economical in emerging nations. Numerous instances of the economical use of "different IT applications" are provided by Zijp's (1994) study. However, occasionally their beneficial effects can only be realized at a high initial and ongoing expense. Moreover, this effect usually arises when the media are employed in conjunction with other innovations that we have examined, suggesting that IT is better viewed as a "force multiplier" that facilitates or amplifies the efficacy of other innovations and traditional extension techniques rather than as an isolated innovation (Antholt, 1994). A further challenge to making generalizations about how IT affects general problems is the variety of media that can be taken into account for various extensions. AED (1985) discovered that the message, target audience, and social context all influence how appropriate certain media are. Print medium is more suitable for timely information reminders, but radio and television are better for rapidly reaching large audiences with relatively simple ideas. The most effective methods for imparting knowledge and boosting information trustworthiness are interpersonal communication techniques like group discussions, extension agents, and demonstrations. Drawing from Lionberger's (1968) adoption process model, Campbell, and Barker suggest:

1. Mass media and popular theatre to supply fresh or supplementary data during the "consciousness" phase,
2. "Interest" stage knowledge can be expanded by field trips, radio programs, and group gatherings,

3. Demonstrations of results and methods, as well as farmer exchanges to enhance proficiency in the "evaluation" stage,
4. Inducing behavioral change during the "trial" stage through one-on-one meetings, farmer exchanges, on-farm trials, and method demonstrations, and
5. At order to solidify attitudinal change at the "adoption" stage, recognition programs, competitions, and integrating methods into farming systems are utilized.

"Interconnectivity" refers to the process of enabling people to connect with each other through the use of appropriate information and communications technology (IT) (Zijp, 1997). It will need a significant amount of dedication and some drastic perspective adjustments to realize its full potential. One shift is to shift more towards a pluralist, cross-sectoral, systems approach and less towards our reductionist, sectoral orientation. Thus, institutional pluralism and decentralization are intimately tied to interconnectedness. Tele-cottages, also known as community communication centers, are prime examples of the new collaborations that are developing for local communication, education, and information access in rural areas. They are just as varied as the communities they serve, however after their initial start-up, many manage to finance themselves, and they all relinquish authority and accountability in order to concentrate on the informational, instructional, and organizational needs of the community. After the first tele cottage opened in 1985, 50 more were operational in Sweden within five years. The objective is to mitigate the negative effects of remote locations by giving remote communities access to computer-aided services, fax, email, databases, online tutorials, open universities, and village hall facilities. Wete (1991) identified a few restrictions:

1. Information on its own is not enough to bring about social change.
2. Information sharing has the potential to increase the wealth gap rather than narrow it.
3. Without the government's commitment to change, demonstrated by its supply of budgetary assistance, supportive policies, and supplementary services, communications technology (CT) cannot have an impact.
4. The majority of underdeveloped nations lack the financial means to purchase CT hardware, which casts question on the benefit/cost ratio of various CT applications.

The limitations of IT/CT have much to do with the fact that breakthroughs in this area do not stand alone when it comes to solving general problems. Therefore, IT cannot solve the cause-and-effect relationship problem on its own. "Garbage in, garbage out" is applicable, and excessive reliance on IT should not offer false or biassed information the appearance of being true. IT cannot completely replace in-person interactions between farmers and extension personnel in terms of coverage. Additionally, certain IT programs include biases towards urban areas and other things (such prejudice against underprivileged people and traditional cultures). However, more funding for numerous IT applications seems to be a wise social and economic decision that should be supported by the public sector (Zijp, 1997).

Chapter 5

AGRICULTURAL EXTENSION GUIDELINE

Contents

5.1. OVERVIEW

Extension is primarily a way of introducing new ideas and expertise into rural regions with the goal of improving the lives of farmers and their families. Thus, extension is really important.

Farmers wouldn't have access to the assistance and resources needed to advance their farming and other profitable endeavors without agricultural extension. It is easier to comprehend the crucial significance of extension when its three primary components are taken into account.

5.2. WHAT IS THE MEANING OF EXTENSION?

An informal educational approach aimed at the rural populace is called extension. This procedure offers guidance and knowledge to empower individuals in overcoming their challenges. Extension aims to elevate the standard of living for farm families, enhancing both the productivity and efficiency of their agricultural endeavors.

5.2.1. What is the Objective of Extension?

The aim of extension is to alter farmers' perspectives regarding their challenges. Extension focuses on enhancing both the material and financial aspects, as well as fostering personal growth among the rural population. Therefore, extension agents have conversations with rural residents to assist them better understand their issues and determine how to address them.

5.2.2. Basic Principles of Extension

The way agricultural extension work is conducted is guided by a few fundamental ideas. These principles differ with respect to the kind of community in which extension education is carried on. Extension service is based on a set of guiding concepts known as extension principles, which are rules for carrying out extension work. The guiding ideas are:

5.2.2.1. Extension Should Start Where the People Are

According to Williams et al. (1984), extension should operate at the people's level, which is determined by their preparedness, interest, knowledge, and comprehension.

It is crucial for extension workers to be aware of the circumstances in order to help individuals go to greater aspiration levels. It entails having firsthand experience with the local environment and circumstances as well as knowledge of the social structure, customs, habits, and economic situation of the populace. Therefore, the first principle suggests that in order to be accepted by farmers, new ideas must be connected to their existing knowledge and experiences.

5.2.2.2. Needs and Interests

Extension need to be grounded in the interests and demands of the populace, which are intimately tied to enhancing their standard of living by raising agricultural output and enhancing their physical surroundings (Williams et al., 1984). Because people's wants and interests fluctuate depending on their cultural background, it is crucial to understand the fundamental needs of those living in rural areas. Only when these two factors are taken into account can extension work. It is also crucial to remember that extension agents need to win over the trust of their audience. This is the case because farmers are perceived as having conservative and fatalistic views. They are hesitant to act without conviction and are cautious to tread on thin ice. For this reason, it's imperative that the extension agent earns their trust (Obibuaku, 1983). They won't be convinced to follow an extension worker's recommendations until they have complete faith in his competence. This is especially true if the extension agent is inexperienced or young and has never worked on a farm. If that's the case, he needs to work on the one or two programs that have the best chance of succeeding first and continue until they do.

5.2.2.3. Assist Farmers to Determine Their Own Problems

Extension should empower farmers by helping them identify their own issues, finding workable solutions, and motivating them to act. This help does not imply that farmers cannot think for themselves or that the extension worker's problems are a mirror image of the farmer's concerns. By starting this, the extension worker will help the farmers identify their issues and give them the impression that they are concerned about their difficulties. The extension worker should not in any way resolve their issues on their behalf by giving solutions, as this would equate to putting his own moral standards on them.

1. **Start with the Felt Needs:** Humans have unfulfilled desires; this is a well-established phenomenon that also applies to farmers. Programs must begin with the needs that the people feel and go on to others that they also require since an extension worker cannot progress with clients unless they are willing to assist themselves. People's desires must be kept in proportion to the amount of work they are able and willing to put in. Communities are not made up of individuals who all desire the same thing at the same time or in the

same way (Obibuaku, 1983). As a result, their ideals, aspirations, and capacities for achieving them, diverge.

2. **The Principle of Co-Operative Work Must Be Pursued to Logical Conclusion:** This is true because the most effective programs are those that are developed in collaboration with local residents and extension personnel. Developing programs in collaboration with the public is a crucial aspect of extension education. When people are involved in the construction of a program, they comprehend it better and are more willing to support it. Moreover, planning involves learning. Participating in program planning teaches people how to collaborate. Collectively generated decisions are more powerful and regarded with greater acceptance than decisions that are imposed or passed by outside parties. The best strategy to assist rural residents in embracing change is to involve them in the planning process, as they are prone to reject change until they realize its advantages.

3. **Extension Workers Should Work with all Members of the family:** In the house and on the job, the family ought to be seen as a cohesive unit (Williams et al., 1984). When working with rural residents, extension workers should consider them as logical individuals capable of making their own decisions, regardless of their race, religion, or political beliefs.

4. **The Principle of the Use of Variety of Teaching Methods Is Another Basic Principle.** In this situation, teaching a segregated learning unit might be thought of as a teaching approach. This is also founded on the ideas that diversity is what makes life interesting and that no single approach can help individuals change in the way that is desired. Therefore, no method exists in a vacuum. This idea implies that people tend to pick up on concepts more quickly the more ways in which a topic is presented and used.

5. **Voluntary:** Since participation in extension programs is voluntary in African rural communities, these programs must be tailored to the unique requirements of participants. There are large differences in age, sex, education, attitudes, needs, interests, and economic and social values among participants in extension programs. As a result, programs need to be engaging and customized to the interests and requirements of the various groups.

6. **Maximum Opportunity:** Extension agents and staff members should collaborate to establish programs and offer as many opportunities for individuals to work on them as possible. While

collaborative program design is important, complete participation is another factor that is essential to the programs' ultimate success. When farmers are allowed to practice what they helped to establish as much as possible, they are completely satisfied. The engagement principle is psychologically valid since people never show interest in programs they did not contribute to developing.

7. **Extension Workers Should Take Advantage of Any Existing Local Groups to Involve the People in Extension Programs**: Since local leaders are influence brokers and the extension agent cannot succeed in his programs if he cannot collaborate with them, rural residents are more likely to listen to them than to the extension agents. If innovation is to be successfully embraced by the community, this rule needs to be closely followed. Extension agents are able to provide their services throughout a large region due to the presence of these local leaders. Ethiopia is home to a wide range of associations and groupings, including kebele administrative councils and farmers' cooperative societies or unions. A greater effort should be made to involve all of these groups in extension activities.

8. **Definite Purpose:** In order for the program to fulfil its intended function, the subject matter addressed in the extension must be precise and have a clear goal. The extension program's content is referred to as subject matter here since it needs to be applicable to rural residents' lives and beneficial to them. The program's material must be delivered at the precise moment when the audience most needs it. This is because, in the absence of opportunities to apply what has been acquired, retention rapidly declines. Therefore, the material covered must be feasible given the time allotted, the clientele's financial and physical resources, their social standing, and their capacity for learning.

9. **The Principle of Constant Evaluation Must Be Followed:** It should regularly evaluate its work in light of the current and evolving circumstances to determine whether the goals are being met. Extension agents are required to make a never-ending list of decisions and then act in accordance with what they believe to be those decisions' directives. Similar to this, it becomes more difficult to remain objective about a practice's shortcomings and to effect the necessary changes the longer it has been in place. Thus, regular evaluation will be quite helpful in achieving these goals.

10. **The Principle of Professionalism Should Be Followed.** Thus, extension agents ought to collaborate with specialists in the field

who can effectively market their initiatives to their target audience. Thus, credibility is crucial in this situation. It should give its employees ongoing opportunities for professional development, further training, and advancement.

11. **Learning Is a Gradual Process and Therefore Results Must Not Be Expected Too Soon:** Empirical data has demonstrated that before students can react to new concepts, they need to be introduced to them in a number of contexts and over time. People in rural areas don't study at the same pace, thus they shouldn't be hurried. When expecting the rural population to perform fundamental tasks, this idea needs to be kept in mind. As a result, the adoption rate should be viewed as a progressive process.
12. **Adult Learning Remains High Throughout Life:** Adults have developed a variety of views, behavior, and experiences throughout years. Their behavior and attitudes typically shift very gradually. If advancement is to be achieved, a lot of these must be altered. Therefore, in order to maintain these attitudes and behavior, the Extension workers must employ every tactic at their disposal. An even closer principle to the one just mentioned is that extension serves an educational purpose by helping people weigh the pros and cons of the options that are presented to them. Activities related to supplies should not be handled by extension personnel. It's possible that the farmers anticipated receiving the necessary fungicides, fertilizers, and planting supplies from the extension personnel. This essentially goes against its intended purpose of educating.
13. **Extension Workers Should Promote the Use and Development of Volunteer Leaders:** Extension agents can reach a large audience and inform them of the need for change through this forum. As a result, this principle views volunteer leaders as extension's loudspeakers. Most of the scheduled programs won't happen if the volunteer leader isn't used.
14. **Extension Should Be Based on Facts and Knowledge:** By closely collaborating with the farmer and researchers, this principle can be realized. Thus, in this sense, extension will be viewed as a bridge or a middleman between the farmer and the researcher.

5.3. THE EXTENSION AGENT

The extension agent, who plays a crucial role in all extension activities, is the foundation of the entire extension process. It makes no difference how creative the extension technique is or how abundant the supply of inputs and resources for extension activity is if the extension agent is unable to react to a given scenario and function successfully. In fact, an extension program's success or failure is frequently determined by how well the extension agent performs. The extension agent must interact with individuals in a multitude of ways. It's frequently a close relationship that requires a great deal of delicacy and creativity. It is inevitable for the agent to deal with individuals whose situations diverge from his own. Working with farmers, many of whom have no formal schooling and may lead completely different lives from his, he is an educated and skilled professional. The agent essentially meddles in the lives of the farmers in a specific area as part of his extension operation. The extension agent acts as a change agent, stepping in to enhance the lives of farmers and their families through intervention. This is not a simple task, and there are several problems with this intervention. In this situation, it's important to think about the fundamental function that an agent plays in bringing about change in a rural area as well as the knowledge and abilities that an agent would need to bring to bear in this capacity.

5.3.1. The Role of the Agent

There aren't any universally applicable models of an agent's role. An agent needs to evaluate every scenario on its own and take on a role or position that best fits it. Regarding the role of the extension agent in bringing about change among farmers, opinions do differ greatly. A variety of comments regarding the role of the agent that are taken from global extension practices can be analyzed in order to demonstrate this spectrum of viewpoints.

- In order to help people feel satisfied and proud of their accomplishments, an extension agent attempts to identify and pique people's interest in their problems, teach them how to solve these difficulties, and convince them to act on his advice.
- The main goal of a change agent is to bring about a change in behavior, attitudes, and social structures.
- Change agents are intermediaries between the public and the government that have multiple functions.
- A person who initiates a change process after determining that specific adjustments are required for rural culture is known as a change agent.
- An activist's primary responsibility as a change agent is to assist others in creating their own organizations so they may address issues as they arise.

- A professional who steers innovation and decision-making in the direction that the change agency deems desirable is known as a change agent.

The aforementioned quotes encapsulate the essence of the diverse perspectives that exist regarding the function of the extension or change agent. It is not implied that one statement is more significant or legitimate than another. They are just meant to highlight the complexity of the agent's job and the necessity for the agent to give changes considerable thought before taking action. Still, a lot of the concepts around the agent's work fall into two fairly general categories. It is possible to consider the agent to have two primary, albeit distinct, areas of responsibility.

- **Knowledge/Communication/Innovation:** It is the responsibility of the extension agent to furnish farmers with the necessary knowledge and information to make informed decisions regarding a specific innovation and to convey this information to them. In this capacity, the agent is viewed as a teacher guiding farmers in the application of knowledge and as a conduit for knowledge, most of it technical in nature. The technical know-how and data that the agent needs to convey to the farmers are given to him together with official training for this role. The agent's work in this capacity is typically very regimented and grounded in the policies and rural development initiatives currently in place from the government.
- **Educator/Facilitator/Catalyst:** The agent is more closely related to the personal growth of the farmers in his capacity as an educator, catalyst, or facilitator than he is to the knowledge and communication component of his work. The agent's primary focus is on assisting farmers in developing self-confidence, organizing themselves, and initiating participation in extension activities, rather than on particular programs or goals. The agent's primary responsibility is to actively support and encourage farmers to take the initiative and start solving their own difficulties.

The broad range of terms employed in extension around the globe to characterize the function of the extension agent can be used to sum up this discussion. The purpose of the list is not to imply that the agent needs to possess each of these qualities. Nonetheless, it highlights the significance of the agent in global extension practice and the variety of ways in which this position can be interpreted (Table 5.1):

Table 5.1. *List Showing the Diversity of an Agent Roles*

Teacher	Facilitator	Organizer	Arbitrator
Educator	Broker	Administrator	Advocate
Leader	Consultant	Enabler	Catalyst
Communicator	Intermediary	Activist	Friend
Motivator	Listener	Provider	Stimulator

Source: FAO.

5.4. KNOWLEDGE AND PERSONAL SKILLS

The kinds of information an agent needs to possess and the soft skills he needs to perform his duties well are two more crucial matters. Once more, there is a great deal of variation in extension practice around the globe regarding these two challenges, which reflects the range of circumstances in which agents operate. It will be examined how the role an agent plays affects the knowledge and soft skills needed by the agent.

5.4.1. Knowledge

Extension training is built around four key areas of knowledge that are critical for the extension agent.

1. **Technical:** The agent needs to be well-versed in the technical facets of his job and possess a solid understanding of the fundamentals of the agricultural system he works in.
2. **Rural Life:** This comprises sociological and anthropological research into the customs, behavior, culture, and values of the rural community in which he works.
3. **Policy:** The agent should be knowledgeable about government legislation, institutional policies that impact rural communities, development initiatives, loan programs, and administrative and bureaucratic processes.
4. **Adult Education:** Since extension involves education, the agent needs to be knowledgeable about group dynamics and the major theories of adult education. As well as with methods for encouraging farmer involvement in extension initiatives.

It should be mentioned that, if the agent is given access to the comprehensive material found in each of the four knowledge domains, the training program can address these four areas.

5.4.2. Personal Skills

Determining the personal skills an agent needs for extension work and training them in these areas are more challenging. The domain of personal talents is more difficult to address through particular knowledge and data, and it frequently pertains to abilities that an agent possesses or lacks. Although many different kinds of talents have been proposed, these have been compiled into a list of the primary skill areas that an extension agent must possess.

1. **Organization and Planning:** The extension agent needs to be able to successfully manage and oversee the operations of an extension office, as well as plan and coordinate the implementation of extension activity.
2. **Communication:** Above all, the foundation of every extension activity is effective verbal and nonverbal communication, which is a talent that an extension agent must possess.
3. **Analysis and Diagnosis:** The extension agent must be able to examine situations which confront him, recognize, and understand the problems that exist and propose courses of action.
4. **Leadership:** In addition to leading by example and taking the initiative to start activities, the extension agent should instill confidence and trust in the farmers he works with.
5. **Initiative:** The extension agent might frequently have to work alone and without supervision. He needs to be self-assured and self-starting enough to accomplish this without needing direction or assistance from his superiors.

5.4.3. Personal Qualities

It's common to talk about the character traits that an effective extension worker must possess. Even if these characteristics are more elusive, it is nevertheless important to consider them while choosing extension agents. Before assigning an agent to an extension post, it's critical to determine whether or not they exhibit certain required personal attributes. The following character traits are necessary for an extension agent:

1. Dedication to extension work and the willingness to work hard and determinedly to start some extension initiatives, even while working in remote rural locations.
2. Dependability in executing extension activities and cultivating relationships with farmers. An extension worker needs to be trusted by his top officers to complete his work without direct supervision and gain the trust of the local farmers in his guidance and assistance.

3. Modest in his interactions with the farmers. The agent has to regard the farmers as individuals with information and ideas to share, and they should be considerate of their needs and desires.
4. Assurance in his own skills and a strong will to succeed. An extension agent must have the bravery and self-assurance to work alone, unsupervised, for extended periods of time.

The information, abilities, and personal traits listed above are not all-inclusive. They are meant to demonstrate the extremely hard nature of the work and serve as a guide for choosing and training extension agents, not as a checklist by which one may assess a person's suitability for doing extension work.

5.4.4. Public Speaking

The agent will need to regularly practice public speaking as a skill. One of the agent's main responsibilities is communication, which invariably entails giving a speech in public to introduce a novel concept, carry out a demonstration, or participate in a general community discussion. When done well, public speaking may be a very beneficial and successful communication tool. The extension agent has the chance to showcase his enthusiasm and technical expertise during a speech in public. Certain people are born public speakers; they effortlessly and expertly organize their content and deliver it in a way that is both captivating and understandable to the audience.

However, an extension agent's efforts can be in vain if he is ill-prepared for his speech and gives a terrible performance in an awkward setting. If an extension agent prepares effectively, public speaking is a valuable skill that he can utilized to his full potential. Most significantly, a talk ought to seem well-planned and have a smooth flow. Nothing turns off an audience more than a speaker who is disorganized, forgets important details, or speaks for an excessive amount of time. An extension agent who is well-prepared and delivers their message with confidence can leave a lasting impression on farmers. Agents may do a lot to enhance their performance, even though not all of them will naturally be good public speakers. With application and practice, the majority of agents will be able to build an effective public speaking strategy. For the extension agent, the following list of guidelines provides helpful advice on public speaking.

1. **Preparation:** No matter how insignificant the occasion may seem, the agent must prepare himself in advance for a public speaking engagement, just like with any extension operations. In order to prepare for the presentation, one must verify the facts, numbers, and other information that will be presented, arrange the information

logically, and prepare any accompanying audiovisual materials. Verifying the location of the talk and accounting for the needs, interests, and background knowledge of the anticipated audience are also helpful. Some agents believe that practicing the talk—or at least the main points—in advance is helpful. The agent should write the key points in order for reference in large font on paper or on small cards rather than attempting to memories the entire lecture. Another tactic is to write the key ideas on a page that can be revealed one slide at a time as the speech goes on. This accomplishes two goals: it aids the agent in presenting and gives the audience a synopsis of the key points.

2. **Content:** A talk's substance should be carefully considered, both in terms of the language and terminology used and the make sense of the order in which things should be said. The talk's topic should be introduced, along with a brief synopsis of it. The primary points that the agent wants to make should be covered in the talk's main body. He should avoid packing too much information into a speech; a succinct, well-planned speech will make a considerably bigger impact than a long, meandering presentation. Additionally, it's critical to use caution when choosing words and phrases and, whenever feasible, modify them to fit the specific situation. In a similar vein, it's critical to refrain from using jargon or extremely sophisticated technical language as this could confuse the audience.

3. **Delivery:** This is the most important aspect of public speaking and needs to be handled carefully. In delivering, confidence is crucial. An agent is likely to deliver a strong speak if he is well-prepared, knows what he wants to say, and delivers it with clarity and impact. It is beneficial to talk to or with an audience rather than at them, and to adopt a welcoming tone. Under no circumstances should an agent talk down to or belittle the intellect of farmers. An crucial aspect of a talk is body language and posture. While some movements are useful for emphasizing a point, excessive fidgeting, gesturing, or other theatrical gestures can frequently divert listeners. The audience's faces should be continuously observed by the agent, who should note any expressions of interest, boredom, or dissatisfaction. Talks should be two-way conversations in which the agent is aware of the impact he is having and ready to respond favorably. Always make sure the agent's voice is loud and clear enough for everyone in the room to hear. He should make an effort to remain composed and refrain from apologizing beforehand for any mistakes. It's crucial

to approach giving a discussion with positivity. Lastly, the speech should last no more than 15 to 20 minutes. Farmers are unlikely to focus for longer than that, thus the content should be limited to the allotted time.

4. **Questions and Discussion:** It is important to inform the audience in advance of the question and answer period that will follow the talk so that they may make necessary preparations. In addition to encouraging audience members to bring up issues, the agent should be ready to start a conversation. It is best to avoid having just a Q&A session because this will stifle real discussion and lessen the talk's instructional value. In addition, the conversation shouldn't go on for too long or be controlled by a small number of farmers. To encourage a two-way conversation, it is also permissible for the agent to pose questions to the audience.

5. **Report Writing:** Reports are a common and significant task for extension agents; in fact, the quantity and frequency of reports that they must create is a common and significant limitation on their job. These reports are usually required by an agent's supervisor and can be somewhat time-consuming. Although producing reports is a crucial part of extension work, the agent should take responsibility for it and take all reasonable steps to prevent becoming overburdened by them or letting the responsibilities of report writing get in the way of doing actual extension work. Report writing is a talent that the extension agent can learn and employ, much like public speaking. The following general tips should be kept in mind when preparing a report. Make sure that all of the data and information that will be included in the report are easily accessible. Decide on the report's overall topic, structure, and presentation style in advance. Arrange the information logically, starting with the report's goal and moving on to the main points and closing with a few notes. Remain succinct! A report that is succinct, clear, and well-organized is significantly more beneficial and efficient than one that is long and meandering, regardless of the agent's level of operation. After the report has been produced, review it to make sure it is clear, organized, and simple to read. An extension agent who is organized and can write reports fast will find it much easier to complete their tasks than an agent who is disorganized and prepares and writes reports on the spot. Writing reports can take a lot of time, so the extension agent will want to cut down on how much time they spend on this administrative task.

5.5. THE USE OF LOCAL LEADERS

An effective extension agent will constantly look to gain the cooperation of nearby farmers in his activity. With responsibility for thousands of farming families, there will only be a small number of professionally educated extension agents in any given region under any given extension organization. Extension agents should look for and recruit the help of locals who possess leadership abilities or influence in the community as a solution. An extension agent can benefit greatly from the support of local leaders in many different ways. In addition to helping to set up local extension groups, they can take on specific responsibilities when the agent is not present, actively promote new ideas and practices by showcasing them in their fields, and function as a general point of contact for the farmers and the agent. The extension agent will be able to contact considerably more farmers by enlisting their assistance than he could on his own. Collaboration with local leaders fosters stronger relationships with farmers in the area and increases their trust in the extension service as well as their readiness to take part in extension-related activities.

5.5.1. Formal and Informal Local Leaders

Extension agents will collaborate with official and informal leaders on a local level. Formal leaders are defined as members of the community who occupy formal positions within the administrative and bureaucratic framework in Chapter 3. Representatives or agents of government departments, traditional chiefs or headmen, educators, religious leaders, political figures, and representatives of local institutions (like cooperatives) are a few examples of these leaders. It is the extension agent's responsibility to make an effort to engage these official leaders in his activities and go over relevant aspects of his job with them. He can establish a solid foundation for his extension efforts in the region if he can get their overall support for his program. It goes without saying that the extension agent ought to invite these official leaders to any public extension events or meetings.

The extension agent will collaborate with local informal leaders to carry out his fieldwork and extension program. Prominent farmers in their community who exhibit traits and skills that an extension agent could find useful are examples of informal leaders. In a rural area, there will be informal leaders, and the extension agent should be able to identify them via thorough observation and questioning. The agent will frequently be able to identify the important farmers whose support will be crucial in promoting extension efforts by speaking with other farmers and finding out who they consider to be the natural leaders in the area.

5.5.2. Selection of Local Leaders

Extensive global extension experience has provided insights into the traits and attributes that agents should consider identifying in farmers who have the potential to become effective local leaders. It goes without saying that an extension agent shouldn't hurriedly ask a nearby farmer to take the lead on extension initiatives. He needs to think about his decision. The traits to seek for in local leaders are exemplified by the next two lists.

1. Initiative to assume the initiative and inspire others. intellect to recognize problems and comprehend issues. Industry and vigor, to collaborate selflessly with fellow farmers. power over others as well as the capacity for instruction and persuasion. honesty as well as accountability.
2. Knowledge about contemporary agricultural techniques and farming. knowledgeable and literate. dependable and consistent in attending extension events. inventive and open to experimenting with new concepts. admired and trusted by his farming colleagues.

Although the aforementioned lists are not intended to be check-lists, they do highlight the significance of selection and the attention that the agent must take in ascertaining the attributes that are significant to him. Next, he needs to look for local farmers who possess these attributes.

5.5.3. Working with Local Leaders

The above-mentioned traits should be carefully developed by the extension agent. It will be crucial for him to maintain his personal relationship with local leaders, and he should make it a point to be accessible to them whenever they need help. The agent should be aware of four key elements while collaborating with local leaders.

1. Notify: Provide local leaders with extension material and updates on proposed new programs and activities.
2. Visit: Pay them a visit as frequently as required to make sure they're not abandoned or left alone. To help the leader incorporate the visits into his own work schedule, try to schedule them on a regular basis.
3. Instruct: Educate the local leaders on areas of extension work that they might not be familiar with. Formal training sessions can be organized to teach the leaders new techniques, demonstration techniques, and farmers' meeting procedures.
4. Motivate: Motivate the leaders in your community to take charge and start acting independently. The extension agent's chances of

having an influence in the community will increase with their ability to become more well-known and efficient.

When an extension agent can enlist the help of a group of capable local leaders, he has access to an enormous resource that will help him get extension work started in that region more quickly than if he had to work alone and without assistance.

5.5.4. Problems of Working with Local Leaders

The extension service can greatly benefit from collaborating with local leaders, but there are a few possible issues that the agent should be aware of. The agent needs to be very aware of his own interactions with the local leaders as well as their field-level performance. Few issues will come up if the leaders are properly chosen and managed, but it would be prudent for the extension agent to be on the lookout for the following possible issues.

1. The problem of favoritism may emerge if the agent devotes excessive time or effort to one or more of the local leaders.
2. As contact farmers, the local leaders are supposed to share the information they have learnt from the agent. It is not always the case that information is shared from the leader to the other farmers, thus it is very important for the agent to make sure that it does. The agent will need to look into the reasons if the leader isn't acting as a contact farmer.
3. Certain local leaders may exhibit a conceited and overbearing demeanor, exploiting their privileged relationship with the extension agent for personal benefit.
4. It's possible that some local leaders are less competent than others, make mistakes, and counsel other farmers incorrectly. It is imperative for an agent to make sure a leader is well equipped before assigning him to oversee extension activities.

The aforementioned instances of collaborating with local leaders are provided not to imply that such situations will invariably arise, but rather to prompt the extension agent to exercise caution in his use of local leaders and to remain prepared to address any issues that may come up.

There wouldn't be any local extension services accessible to farmers without an extension agent in the field to advise, direct, and oversee these operations. The extension agent is the essential component of the entire extension process.

The crucial elements of this procedure are the agent's function and relationship with the farmers; things don't always go as planned. Global extension experience has emphasized the agent's pivotal role and brought to light certain characteristics that set a successful extension agent apart from a less successful one. A productive agent

1. Invests time in helping farmers improve their own abilities and qualities rather than focusing only on extension initiatives,
2. Gets out to visit and meet farmers and does not become an office bureaucrat,
3. Does not take a paternalistic stance towards farmers and instead promotes local initiative and self-reliance,
4. Plans for the long-term development of his area, and does not only seek quick results.

Once more, the problems listed above are merely illustrations of what could happen in any extension region. In reality, the majority of extension agents are devoted individuals who labor in challenging circumstances and frequently without any assistance. An extension agent's job requires specific traits including hard work, humility, and dedication. Extension services should make sure an agent is well-prepared before starting any extension activities.

5.6. AGRICULTURE EXTENSION PLANNING AND M&E

It has been emphasized that meticulous planning is necessary for any extension activity to be successful. Every demonstration, open forum, and film screening is a component of a larger extension program that involves farmers and extension agents together to further the region's agricultural growth. No extension activity is organized in isolation. Four components make up a written statement that is an extension program:

1. Objectives which the agent expects to be achieved in the area within a specified period of time. This will often be a one-year period, to enable the agent to review the program at the start of each farming year.
2. Means of achieving these objectives.
3. Resources that are needed to fulfil the program.
4. Work plan indicating the schedule of extension activities that will lead to the fulfilment of the program objectives.

Local farmers, the extension agent, his senior extension officers, and other rural development organizations can all benefit from an extension program with well-defined objectives. It demonstrates to farmers what to anticipate from the extension service as well as the efficacy of the agent. The program gives the agent a solid foundation for organizing extension operations on a weekly and monthly basis and for foreseeing well in advance what resources would be required. Programs can be used by senior extension officers to evaluate agents' work, provide suggestions for enhancement, and support requests for more employees, supplies, and money. Additionally, the program assists other agencies in coordinating their operations with the agent's. Programs can only be written and made available to all parties involved if they are to be utilized in these manner. When an agent is brought into the region due to personnel changes, a documented program might also be helpful. The program allows the new agent to pick up where his predecessor left off, guaranteeing that the operations continue. Every organization engaged in agricultural growth has its own planning processes, which can differ greatly from one another. They may vary, in instance, in how much planning is done at the local or national levels. The effectiveness of extension initiatives depends on careful planning, monitoring, and evaluation of the activities, results, and implications of the program. In response to regional needs, the extension department is in charge of offering farmers superior agricultural services. There are two options for the kind of service and how it is delivered:

1. Extension events planned in advance which form the Annual Extension Plan.
2. Extension events planned in response to day-to-day interaction with farmers as follow up activities.

All extension plans are prepared in either top-down or bottom-up approach (Figure 5.1).

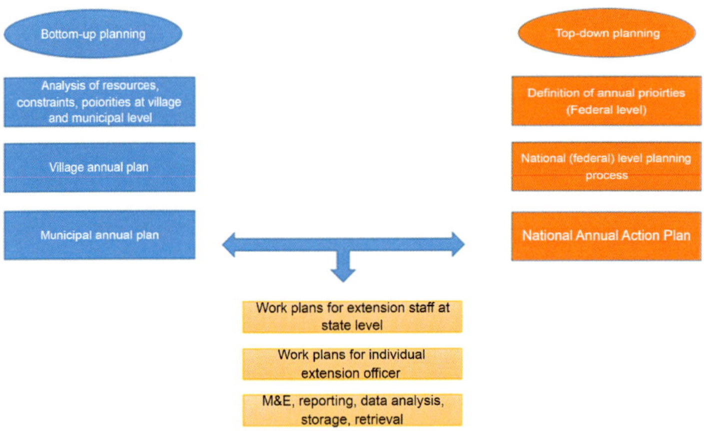

Figure 5.1. *Approaches to planning extension.*

Source: FAO.

Both of these planning strategies should be used in successful extension efforts. The agent plans his local activities within the framework provided by national policies and programs; these also set priorities that he must adhere to. The agent will prioritize these crops in his own program if increasing the production of arable food crops rather than livestock products is a national objective. Additionally, national programs will provide resources and funding for specific types of activities, which will affect the agent's local planning choices. However, farmers must take voluntary effort to enhance agriculture if they are to raise their own standards of living and productivity. Therefore, local requirements serve as the driving force behind agricultural development and should be considered while designing local extension initiatives. The agent must nonetheless create a plan that will allow him to carry out national policy and directives within his jurisdiction, even if those policies and directives restrict his freedom of choice. It is therefore important for the agent to consider both local and national criteria while designing his extension program. He should be aware of national goals, but he should also collaborate with the local community to ensure that the program that is developed represents their needs and their desires for the community and their farms. This local participation in the planning phase is a crucial component of the extension education process. It encourages a thorough examination of farming issues and supports the development of drive and self-assurance in utilizing regional resources to address them. When organizing extension projects, agents collaborate with official committees at the local level in certain nations. Regardless of how this local involvement is accomplished, the agent needs to ensure that individuals involved are genuinely able to represent the interests and viewpoints of all local organizations. Committees tend to favour programs that align with the objectives of larger-scale, progressive farmers because they frequently comprise a significant number of these individuals.

5.6.1. Extension Program Planning Guide

Five separate stages can be distinguished, regardless of the specific procedures for program planning that the extension organization specifies.

1. **Analyze** the present situation.
2. **Set objectives** for the extension program.
3. **Develop the program** by identifying what needs to be done to achieve the objectives, and then prepare a work plan.
4. **Implement** the program by putting the work plan into effect.
5. **Evaluate** the program and its achievements as a basis for planning future programs.

After that, the situation will be reviewed and a new program will be planned, with the goal of building on the successes and learning from the mistakes of the previous one. The many phases of planning an extension program are connected to one another, and the planning process does not always flow smoothly from one to the next. For example, during the situation analysis stage, provisional objectives may be created, but they may later be changed if more information becomes available and a more thorough investigation provides a more comprehensive picture of the circumstances. Goals could need to be adjusted even further if unforeseen issues crop up while the program is being run. Nonetheless, since each of these five actions may be divided into smaller parts, it is helpful for the agent to think about program planning as involving these five activities. These will all be discussed in greater detail.

1. **Situation Analysis:** Prior to creating an extension program, the current state of affairs needs to be examined. It is necessary to comprehend farming issues and their root causes, as well as to identify the region's natural, human, and other resources. There are three tasks involved in this phase. Every year, a comprehensive scenario analysis is not required. Most of the time, the fundamental information about the region and its inhabitants won't vary significantly from year to year. Nonetheless, the agent should assess the fundamental data annually and determine which sections require updating.

2. **Collecting Facts:** The agent will require a great variety of information regarding the local population, farming practices, natural resources, and amenities available for local growth. For example, information on social structure and local culture, farming systems, education, and literacy levels, farm sizes, local channels of communication, transportation facilities, local credit systems, marketing, health, and nutrition levels, and crops and livestock are all necessary for the agent to have a thorough understanding of the situation with which he is dealing. There are numerous places from which these details might be gathered. A wealth of helpful background data can be found in reports from past projects, social surveys, farm management studies, land-use surveys, and soil classification surveys. A large portion of the information the agent needs will be available to him if he maintains thorough records of the farms in his region. He can also get a lot of information from local authorities and farmers. In public gatherings, group conversations, and private interactions with farmers, the agent should pay attention, probe, and gradually develop a deeper comprehension of the region's social, agricultural, and economic characteristics. It is advisable to leave

sophisticated questionnaires and statistical analysis for expert researchers when conducting in-depth social and economic surveys. However, straightforward questionnaires can be employed in this fact-finding process, and it is beneficial if the extension organization gives the agent a uniform list of questions or information to refer to. Farmers can be involved in the planning process from the beginning by conducting some of the fact-finding themselves. For example, they can use basic questionnaires and check-lists to compile a list of the region's natural resources.

3. **Analyzing Facts:** Data cannot speak for itself. It is important to consider the reasons behind events as they unfold. In the event that farmers indicate a decrease in yields over the past few years, the agent needs to investigate further for other data that could provide insight. Is it a result of inadequate seed, decreased soil fertility, or little rainfall? The agent must also distinguish between opinion and conjecture and actuality. He might receive contradicting information from two sources, and he will have to decide which is more trustworthy.

4. **Identifying Problems and Potential:** At this point, it ought to be feasible to identify the primary issues that local farmers are dealing with and the possibility for agricultural advancement. This is where the technical expertise of the agent is crucial. Farmers may be aware of their own issues, but the agent can offer his unique perspective on regional issues derived from a deeper comprehension of science in agriculture. He will be able to go into greater detail when describing issues and connecting them to procedures that farmers might not be aware are related in any manner. Owing to his education and background, he will also be better equipped to offer recommendations on how the region's resources might be put to better use. Throughout the situation analysis, the agent should refrain from either fully depending on his own knowledge to interpret facts and identify issues, or fully delegating to farmers the task of defining local requirements and opportunities for change. To achieve a complete understanding, it should be a collaborative endeavor with agents and farmers contributing their individual experiences and insights. The agent faces the risk of misinterpreting information, wasting time on analysis, and, quite likely, not getting the full support of farmers for the program if farmers are not completely involved in these activities.

5. **Setting Objectives:** Following an analysis of the current state of affairs, judgements can be taken regarding the modifications that

ought to be implemented via an extension program. The two most important considerations are how local potential will be utilized and how local problems will be resolved. It will be necessary to define three stages of distinct, achievable goals in order to find solutions.

6. **Finding Solutions:** The agent should differentiate between technical solutions—better inputs or straightforward adjustments to husbandry practices—and institutional solutions—better credit and marketing channels, for example—while searching for answers to local issues. Reforming institutions may call for higher levels of government and other agencies to take action. There might not be much that can be done locally in isolation, even though the agent should definitely recommend such solutions to those who are accountable. Some places to get inspiration for realizing the potential of a place are:

- the agent's own technical knowledge;
- farmers and agents from other areas who have tackled similar problems successfully;
- applied research which tests new ideas under farm conditions;
- national priorities and directives;
- projects which make funds available for particular activities.

7. **Selecting Solutions:** Agents and farmers should make sure that suggested solutions are the following when choosing from the variety of fixes and enhancements that are available:

- Acceptable to farmers in the area.
- Technically sound and tested by research and experience elsewhere.
- Consistent with national policy, and with the local activities of other agencies.
- Feasible within the time and with the resources available to farmers and the extension service.
- Within the scope of the agent's ability and job description.

During the duration of the extension program, the agent may discover that certain issues cannot be resolved locally or in a way that is both acceptable and possible.

Legislation, actions by other authorities and at different levels, or additional study may be necessary. When feasible, the agent should mitigate the impact of these issues and serve as a conduit for advocating for reforms to those in a position of authority.

8. **Stating Objectives:** It should now be feasible to specify the goals of the extension initiative. The agent must choose which goals are more important than others, though, since his time and resources may be restricted. He should take into account national priorities as well as the amount and distribution of benefits that result from a specific time and resource input. Whenever feasible, goals should not be stated in general words; instead, they should be stated in terms of amounts and figures. "Establish two groups of dairy farmers who will share new equipment and market their produce jointly" , as well as "Increase the acreage of improved rice varieties from 60 to 120 acres" are more beneficial goals than "Improve dairy farming techniques" and "Increase the use of improved rice seed" . They provide the agent with specific goals to strive for and a benchmark by which the program's efficacy can be assessed at the conclusion of the year. An annual extension program's objectives will outline the goals that must be met by the program's conclusion. These statements can be divided into intermediate actions that must be completed throughout the year to meet the goals of the program. Once more, the agent will need to make decisions by picking the best course of action from a range of options. The agent will effectively be creating a calendar of extension actions for the program term as he dissects each program objective into discrete segments. He will determine what technical information specialists and researchers will need to provide, what extension techniques to employ, what knowledge and skills the farmers will need, and what resources and assistance he will need from his own and other organizations (Table 5.2).

Table 5.2. *Sample Extension Plan*

Problem	**Local Shortage of Staple Food**
Potential	Some local farmers have increased maize yields by 30–40% by using improved seed and animal manure most farmers have plenty of manure but do not use it.
Solutions	A. An Increase size of small farms and introduce labor-saving techniques not feasible: no access to additional land. B. Encourage larger, mechanized farms to grow more food crops not acceptablc: poorer farm families could not buy the food. C. Enable smaller farmers to increase yields feasible, using better varieties and tested husbandry improvements.
Preferred solution C	

Objective	Raise yields by 30% on 20% of the farms in the area in the first year.
Intermediate objectives	50% of farmers will learn of the benefits, and will acquire the skill, of using manure. 50% of farmers will learn the benefits of new varieties. 20% of farmers will plant improved varieties.
Plan of work	Establish demonstration plots on 10 farms. Hold public meetings and film shows in 10 villages to show the benefits of new varieties and improved husbandry. Hold method demonstrations of manuring and correct spacing at the appropriate times. Visit farms during planting season. Hold result demonstrations on selected plots to encourage more farmers to try the new varieties and practices next year.
Support needed	Subject-matter specialist to speak at public meetings, attend demonstrations and answer farmers' questions. Adequate supplies of seeds, at the right time, at local stores. Farm broadcasts to include relevant items at the appropriate time of year.

Source: FAO.

After completing the planning for additional program goals, the agent can combine all of the plans into a single annual work plan. Some of the lower priority goals may need to be abandoned or toned back since he might discover that he is unable to complete everything that each of the individual plans calls for. Although the annual work plan does not outline the agent's daily responsibilities, it should include the start and finish dates of each extension activity as well as the resources required.

9. **Implementation:** The agent completes the tasks outlined in the work plan in order to implement the program. His thorough weekly and monthly plans will consider the issues and advancements from the previous month. For instance, if more farmers than anticipated wish to participate, the schedule for some activities would need to be adjusted, or more method demos might need to be scheduled. An extension program ought to be adaptable enough to let the agent react this manner to situations.

10. **Evaluation:** Throughout the year, the agent will be analyzing and assessing his progress on a regular basis. An in-depth assessment should be conducted at the end of the year, during which the agent determines the extent to which each goal has been accomplished and

the causes of any setbacks. This assessment serves as the foundation for organizing the program for the following year, together with an updated situation analysis.

5.6.2. Evaluating Extension Programs

The process of determining an extension's effectiveness is called evaluation. It involves more than just determining what occurred; it also entails making a determination about what transpired. Was the program's outcome satisfactory enough? Did it turn out as predicted or worse? Was there more that could have been done? Extension initiatives are assessed to:

1. Find out how effectively agents do for the extension organization so that it can determine whether or not they are ready for promotion,
2. Satisfy the government that public money spent on extension is being used effectively, and
3. Let the agent draw lessons from the experience. Time is wasted on evaluations unless the findings have an impact on choices about future extensions.

One way to evaluate a program is to find out if the goals were met. This is a crucial first step that would be made simpler if the program had definite, well-defined goals. If the response is negative, there won't be a solid foundation for future program enhancements. Therefore, it is crucial to enquire as to why events transpired as they did. The agent can only draw lessons from the past once that question has been resolved. As a result, agents ought to enquire about the following software features.

- **Results:** What happened as a result of the- extension program? Were they the results that were expected, and were there any unexpected results?
- **Inputs:** Were all the planned inputs available and, if not, why?

5.6.2.1. Nature of Evaluation

1. **Evaluation Is Not Measurement:** An essential component of extension education is evaluation. Evaluation is required for every facet of extension work. Evaluation is more than just tallying results, which is often donc after a program is carried out. Since extension is an educational process, it is important to assess how the program is managed, what strategies are employed, whether accomplishments are in accordance with the goals, and why certain things happen the way they do.

2. **Evaluation Is Not Exactly Scientific Research:** When we consider assessment to be the process of gathering data to support judgements, conclusions, and decision-making, we find that it shares many similarities with scientific study. However, there is a significant distinction between scientific study and our informal, daily assessment. But rather than being of a different kind, the difference is one of degree. Scientific research might be positioned at one end of the scale and informal everyday judgement at the other. The informal everyday evaluation, self-checking evaluation, do-it-yourself evaluation, extension evaluation studies, and scientific research are the five areas on the scale where there are no clear boundaries.

There are three types of evaluation

1. **Self-Evaluation:** It is expected of every employee to do this on a regular basis. This calls for the kind of self-criticism that extension work so desperately needs. An extension worker has a better chance of developing and maintaining his professional competency if he adopts this self-critical mindset.
2. **Internal Evaluation:** The organization in charge of the program's conception and execution conducted the evaluation. Other techniques for internal evaluation include using employee diaries and reports in a systematic manner, scheduling staff visits to the workplace, using specific questionnaires and proforma for observation and inquiry, etc.
3. **External Evaluation:** An assessment carried out by a group or individual not involved in the operation. The fact that the Program Evaluation Organization, an independent organization, was founded concurrently with the inception of the Indian Community Development Program is one of its strongest points.

Evaluation can also be classified into:

i. concurrent and
ii. ex-post facto evaluation.

5.6.2.2. Levels of Evaluation

Extension programs are evaluated at many levels. It is possible to assess the impact of extension on family incomes, living standards, and agricultural output at the broadest level. One of the main objectives of rural extension is often to raise family living standards, thus it's critical to find out if this has happened. For this form of evaluation, a representative sample of farm families' production and farm income are measured, and the results are compared to past levels. These statistics then show changes that are associated with the program's extension inputs and activities. Extension, however, is not the only factor that raises living standards and productivity; price fluctuations and variations in input availability are just two of the numerous other factors that impact the amount of crop and livestock production.

It is advisable to leave the complicated process of separating the effects of these different factors to specialized research and assessment experts. However, the agent needs to be conscious of these shifts in the economy and constantly assess the extent to which his extension efforts are boosting the incomes of farmers and their families. He ought to see who gains from the extension as well. For example, are the benefits distributed evenly across the farming community as a whole, or do some groups profit more than others? The extension program offers an evaluation at an intermediate level. Here, two enquiries are crucial.

1. Did the extension activities take place in the planned sequence and at the right time?
2. Did these activities lead to the expected results?

The agent should attempt to ascertain the reason if the responses are unfavorable. It's possible that he had unrealistic expectations regarding the number of extension activities he could do or that other organizations did not provide him with the necessary support. For whatever cause, the review process will provide the agent with learning opportunities. In order to guarantee that the required assistance and inputs are given, he ought to be able to create more practical plans in the future. Lastly, each extension activity can be evaluated by the agent. Every extension activity, including speeches, demonstrations, and gatherings with a farmers' organization, has a goal. Whenever feasible, the agent should make an effort to verify if the goal of the action was accomplished in addition to how well it was carried out. Typically, this entails determining whether the extension activity resulted in any modifications to any one or more of the following:

1. Awareness of a particular idea, possibility or problem,
2. Motivation to act in a particular way,
3. Knowledge about new farming practices,
4. Skills needed to adopt a new practice,
5. Behavior by farmers and their families (such as new farming methods), or by an extension group (such as making an application for funds to carry out a group project or the preparation of a formal group constitution).

The agent must gather data at every stage of assessment in order to contrast the pre-activity state with the post-activity one. Prior to evaluation, he will have evaluated the circumstances and determined whether the extended activity was necessary.

For example, he will know how many farmers in the region are aware of, interested in, or have already embraced the specific practice that is to be exhibited while organizing a result demonstration. Asking folks who attend the demonstration about their prior knowledge and experience with the activity will help him get a more accurate appraisal. He can get the data he needs for evaluation by conducting a comparable assessment following the demonstration. More quickly than others, some effects can be evaluated. For instance, the agent can speak with a small group of attendees just after a public meeting to see how well they grasped what he was saying. On the other side, behavioral changes take time to manifest, thus the agent has to wait before examining them. There are various methods for gathering data for village-level appraisal.

1. **Agent's Reports:** Regardless of the requirement for agents to submit a formal report on every extension activity from their extension officers, the agent should take notes on each activity for his own records, focusing on how he conducted the activity and any important points to remember for later.
2. **Supervisors:** An agent finds it difficult to evaluate his performance during an extension activity since, in particular, he is unable to view himself from the perspective of the participating farmers. Thus, receiving helpful criticism from a manager or coworker is beneficial.
3. **Discussions:** Farmers' initial responses will be revealed through casual conversations held with them following the extension activity. Using a tape recorder to capture such conversations is frequently helpful for subsequent transcription and more thorough analysis.
4. **Questionnaires:** When the agent has the time and resources to conduct a more formal evaluation of extension operations, simple checklists and questionnaires can be used. For instance, the agent could compile a list of four or five key information that farmers should be aware of after attending a result demonstration on early planting. The agent can gauge the effect of the demonstration on farmers' knowledge by questioning a sample of farmers about the facts both before and after it takes place.
5. **Observation:** Observation is a reliable source of information regarding changes in farming practices. The agent has the ability to observe if farmers in the region are following his instructions or not.

Numerous extension organizations have official evaluation processes of their own. In certain cases, the agent creates a comprehensive work schedule on a standard form at the beginning of each month, outlining his daily schedule and how it fits within his annual extension program. At the conclusion of the month, the work plan serves as the foundation for review.

1. Did he do all that he planned?
2. Did he encounter any problems that he should take into account in the future?
3. Is he on target in terms of progress toward his annual extension program objectives?

This process may be coupled with a monthly gathering of agents within a certain area to discuss issues and progress in their respective fields. Regardless of the official protocols inside a specific establishment, the agent ought to consider evaluation as a mindset. He needs to become in the habit of questioning what went wrong, why it happened, and how it could be done better in the future. He will keep learning and become better at his extension work in this way.

5.6.3. Extension Evaluation Process

The literature contains a number of evaluation models. Since every extension evaluation procedure must be predicated on certain assumptions, as Bhatnagar (1987) has shown, a highly simplified version of most of these models might be quite practical for assessing extension programs. For instance, if certain inputs are given in the form of a program, certain outputs can be anticipated; if these outputs materialize, the program's aim can be accomplished; and if the goal is accomplished, the development objective is realized. This means that the design of the assessment process must take into account the various quality kinds and levels of adequacy of the input measures, outputs, and their influence on accomplishing the program's goals. The following steps could be included in the evaluation process of an extension program:

5.6.3.1. Formulate Evaluation Objectives

It is necessary to establish and begin the evaluation process with well-defined, sufficient goals in mind. All future endeavors ought to be centered around achieving these goals.

5.6.3.2. Classify Program Objectives

It is expected that every extension program will have precise, well-defined goals when it is developed and put into action. Program objectives must be clearly understood and, if necessary, further broken down into measurable terms. This is because evaluation is essentially a process of determining the extent to which various extension teaching activities were organized and managed and the extent to which they contributed to achieving the goals. This is an important phase since everything that comes after it will be focused on gathering data to support these goals.

5.6.3.3. Identify Indicators

It is necessary to identify specific program beneficiaries, clearly state the behavioral changes expected of them, and specify the kind of learning experiences expected to be provided, along with the level of management to be attained for those learning experiences, in order to identify indicators or the kind of evidence needed to evaluate achievement in relation to specific program objectives. After this is finished, it will be easy to identify particular indicators to gauge the accomplishments.

5.6.3.4. Decide the Kind of Information Needed

Once a program's management and performance indicators have been established, it may be possible to determine the precise data that needs to be gathered. An extension worker must exercise extreme discrimination in determining the type and quantity of information that should be gathered because there is typically more information than he can gather. It might also be necessary to specify the time frame for information collecting.

5.6.3.5. Sampling

Sampling is the process of selecting a comparatively small number of units from a population so that the data gathered from them is representative of the population as a whole. While there are several sampling techniques, stratified sampling procedures might be the most appropriate for extended evaluation studies since they allow for the inclusion of all interested groups and guarantee a sufficient degree of sample variation.

5.6.3.6. Decide the Design of Evaluation

An experimental evaluation design could be the best. By establishing control and treatment groups, this would enable the program's impact to be isolated from external influences. Numerous experimental designs are accessible in the literature and can be employed, including the Solomon four-group design, static groups comparison, one-group pre-test-post-test design, pre-test, post-test control group design, longitudinal study design, etc. But in reality, extension initiatives are rarely managed in a way that permits an experimental evaluation design. It may be feasible to employ an experimental evaluation design in pilot projects. Generally speaking, a survey approach is used. This approach can be applied to program evaluations that are ongoing or ex post facto, meaning they are conducted after the program's duration has ended.

5.6.3.7. Collection and Analysis of Evaluation Evidence

Information can be gathered for evaluations through a variety of techniques, including case studies, mail surveys, in-person interviews, distributed questionnaires, group interviews, systematic field observations, and the methodical analysis of secondary data. The goals of the assessment, the type of information required, the time and resources available, and the type of respondents from whom information is to be collected will all influence the choice of the most appropriate data collection method. Whatever the approach, though, careful consideration must go into creating a particular questionnaire, interview schedule, or data collecting sheet. After the data is gathered, it needs to be carefully tabulated, summarized, and analyzed. It is not necessary to hasten this stage. However, analysis can be completed quickly by using a computer.

5.6.3.8. Interpretation of the Results in a Proper Way

It is quite important because evaluation results can also be overlooked. In order to place the evaluation's findings in the right context, it may be important to discuss the preliminary generalizations that have been reached informally with program planning and implementation officials as well as the interpreters. The evaluation's findings must explicitly list the successes, shortcomings, and necessary future modifications. The results of the review should be documented in writing and distributed to all parties involved.

5.6.4. Special Target Groups in Extension Planning

Rural extension programs should take into account the requirements and interests of the various members of the farming family because they are focused on the entire farming family. The current handbook addresses extension generally, and all rural extension work should be able to benefit from the ideas and methods it offers. Nonetheless, as previously mentioned, each family and community members have unique needs and experience unique challenges, which should be considered in extension initiatives. Three specific groups—women, young people living in rural areas, and those without land—will be briefly discussed here. We'll identify the kinds of challenges and issues that could come up with them in both agricultural and non-agricultural extension work. The fundamental information in the guide is applicable to extension work with all members of the farming community, it should be emphasized once again. Nonetheless, it is believed that these three categories pose unique challenges for expansion, which it would be helpful to pinpoint.

5.6.5. Extension and Rural Women

Agricultural extension services frequently focus more explicitly on farmers—who are typically thought to be men—and their diverse issues with managing and using agricultural resources. On the other hand, non-agricultural extension initiatives usually target women and aim to enhance family resource management or child care within the house. But this separation is common and not necessarily acceptable. Due to the fact that many families do not have a male household member or because women in certain countries are in charge of tending to their own land and crops, many women are independent farmers. Women may perform more than half of the farm work, even in households headed by men. Therefore, it is crucial that agricultural extension collaborate with women as well as men to provide them with the assistance, information, and skills they need to enhance their activities, in addition to any programs specifically created for rural women. Indeed, in the last 10 years, there has been a growing interest in exploring how women fit into the rural development process, realizing the unique contributions that women can and do make, and putting these ideas into action through projects and programs that aim to better the lives of women. Until recently, rural women's issues have received little attention from policymakers or scholars, nor have their unique concerns been adequately understood. Given that most extension agents are men, it's possible that they don't fully comprehend the role that women play in rural communities. To have a better understanding of this perspective, let us first examine the three fundamental responsibilities that rural women play.

1. Economic, both as suppliers of food and other items for the household budget and as workers in commercial enterprises.
2. Domestic, with duties to take care of and oversee the household finances like spouses and housekeepers.
3. Reproductive, in the sense that mothers have duties to care for and raise their children as well as replicate the labor of the family.

Understanding these three primary responsibilities that rural women must play is crucial for extension agents, since they impact women's capacity to engage in outreach initiatives. They also list the many sorts of useful extension support. In rural areas, women typically spend long hours growing the primary food crops for their families. In addition, they oversee the management of the household finances and other economic endeavors like breeding chickens or growing vegetables in an effort to supplement the family's food supply. They are also in charge of preparing and storing the food for the family. Additionally, they become parents and shoulder the majority of the burden for raising and caring for their offspring. Rural women put in a lot of effort and put in long hours, usually for little pay, and they are sometimes overlooked by extension programs. It is imperative that the extension agent makes an effort to comprehend the reasons behind the frequently low level of interaction between his service and rural women. In order to incorporate women in extension activities more, he should first analyses the current state of affairs, identify the barriers that stand in their way, and take those factors into account when organizing these events. Numerous barriers that rural women face and that prevent them from being more involved have been revealed by recent studies on the subject. The following is a summary of these challenges:

1. **Cultural:** Religious beliefs and regional customs are closely related to cultural barriers. Certain communities forbid women from having direct conversations with men who are not in their families. Others have customs that prohibit them from meeting in public, and many explicitly ban women from engaging in activities outside the home.
2. **Domestic:** The weight of household responsibilities severely hinders women's ability to participate more broadly. In addition to managing and caring for the family home, women work full-time jobs that contribute to the domestic economy.
3. **Status:** It is not expected or encouraged for women to actively participate in extension activities since they are typically viewed as less valuable than males. It is very impossible for low-income rural women to rise above their assigned status and participate in development.

It will be challenging for the agent to successfully integrate women into extension activities. However, women's contributions to rural development are so significant that an extension agency needs to collaborate with women as much as possible. Rural women are not inferior to men, and the agent must never think so. They are not, but they do have a distinct set of competencies. The agent should make every effort to deal with men and women simultaneously. He should, for instance, encourage women to participate in meetings and demonstrations. However, women are responsible for distinct tasks, so the agent should focus his extension efforts on these duties. Before starting any projects, the agent must, however, thoroughly research and comprehend the status of women in his extension region, as well as be cognizant of their unique requirements and issues. While women should be included in the agent's general extension efforts to the greatest extent possible, specifically designed programs to strengthen women's responsibilities in rural society will also be required. These initiatives could consist of:

1. Women's clubs and groups are examples of local organizations that should be strengthened, supported, and encouraged in their endeavors.
2. cultivation initiatives that are specifically meant to support women-dominated agricultural tasks, like the cultivation of food crops.
3. initiatives in the field of health care that train women and supply the resources needed for family health care, such as nourishment.
4. Income initiatives like vegetable gardening or handicrafts are made to assist women in augmenting their income.

It is true that, as family heads, males have profited from extension assistance to a larger extent than women have, and that women have seldom been encouraged to participate equally in extension activities. However, it is well acknowledged that women do play a significant role in rural development, and that role should be supported by extension. It is important to recognize the true barriers that women encounter, and extension agents should look for opportunities to divert extension funding towards initiatives that directly benefit women.

5.6.6. Extension and Rural Youth

In many nations, the majority of people reside in rural areas, and a sizable portion of the population is under the age of 18. Consequently, youth comprise a significant share of any rural population. Extension must take action to help these young people get ready for the future since they are the farm families of the future. Alternatively, a great deal of youth will just keep moving into towns and away from rural areas. For these reasons, extension agents must to put out extra effort to engage youth in their work; they ought to speak with students in classrooms and set up field trips for them to observe farm operations. It is intended to include both boys and girls when discussing rural kids. Practice has demonstrated that the boys and girls in question are between the ages of 12 and 18, notwithstanding the lack of precise classifications. These youths are currently the focus of specific extension initiatives in many nations. The Jamaican Youth Corps, the Village Polytechnics in Kenya, and the Youth Voluntary Service in India are a few examples. "Catch them while they are still young," "show concern for the future of young people," and "try to get them started and established in the rural area" are the guiding principles of these programs. The extension worker encounters some unique issues with young people from rural areas. Before starting any youth-oriented extension efforts, the agent must first make an effort to comprehend these issues and how they affect young people's possibilities of improving themselves. Extensive work with young people across the globe has identified the following types of prevalent issues:

1. **Motivation:** Young people lose motivation and frequently lose hope for their own futures when they witness the neglect and backwardness of rural communities.
2. **Training:** Even though a large number of young people have received formal education, many nevertheless lack the skills necessary to support themselves in rural areas. Youth frequently lack the skills and preparation necessary to meet the challenges of rural development.
3. **Involvement:** Young people frequently feel alone and unable to participate in local rural development initiatives. There is no way for them to be heard or to be represented.
4. **Opportunities:** Too few organizations or programs aim to engage youth in rural development efforts or make an effort to reach out to them.

While it is obvious that an agent cannot address every issue right away, he may at least decide which youth extension activities to prioritize in his program. He should establish and keep in touch with the young people in his community, and he should start thinking about creating engaging extension activities for them right away. These events frequently take the shape of youth clubs with specially designed initiatives for their advantage.

5.6.6.1. Clubs for Rural Youth

Clubs have long been utilized as a conduit for youth involvement in rural development, allowing them to suggest projects that will benefit them. The most popular are the 4H Clubs (hands, health, head, and heart), which originated in the US and are currently present in many other nations.

Similar clubs have been established in various nations with the goal of bringing extension into closer contact with rural adolescents, such as the 4K in Kenya and the 4S in Panama. These clubs are crucial as a starting point for bringing young people together because they provide a platform for them to discuss issues and opinions and serve as a foundation upon which to develop. Through them, young people build relationships that will come in handy when they start their own farming families and want support from the extension services. They also get accustomed to working with the services. The extension agent should think carefully about starting a youth club in his community and keep in mind that these organizations serve three primary functions.

1. **Educational:** A club can provide as a structured and informal platform for young people to socialize and prepare for life beyond high school. The club can aim to foster specific skills, leadership abilities, and a general comprehension of the issue of rural development.
2. **Economic:** More specifically, a club can be used to teach young people various aspects of home economics, farm management, and agricultural practice. The club may also take on particular initiatives meant to generate revenue for the young people.
3. **Recreational:** Not every club activity ought to be taken very seriously. Additionally, it need to promote social gatherings and leisure pursuits like sports, day trips, and even dancing. The youth will appreciate the club and regard it as a significant component of their free time in this way.

Before starting a club, the agent should speak with locals and make sure the parents are on board. In addition, he needs to arrange a meeting spot and provide some funding to make it run. Similar problems develop in the agent's work with farmers' groups as they do with clubs. The choice of a club leader and the internal structure of the club are two crucial factors. The club leader is frequently a progressive farmer in the area or possibly a teacher. The leader's job is to assist in directing and encouraging the club's activities; he is not a club member. The leader (or leaders) will oversee the club, assist in project selection, and serve as a general counsellor to the club's participants. Young people will need to be involved in the club's administration as it grows. This could take the shape of a membership committee led by a chairman, treasurer, and secretary. It is imperative that the agent pushes the club to establish an internal structure in order to give discussion and project work some structure.

5.6.6.2. Projects with Rural Youth

Young people can learn to do something more than just listen to speeches or lectures by working on projects, either with the clubs or with rural youth in general. The agent ought to support youth-focused projects and set aside a portion of his local budget for them. These initiatives can be carried out individually or as a club, and they shouldn't start out overly ambitious. Introducing young people to a farmer or other agricultural projects so they may observe a specific activity firsthand is a helpful method to get them started on a project. Local farmers are frequently most open to working together with a group of enthusiastic club members. The agent could also set up presentations by other locals or tours to give them an overview of a project. It is critical that the agent is passionate about working on projects and makes an effort to include young people in the deliberation and selection of projects. Across the world, youth projects that have been successfully completed include home improvement, fish pond farming, vegetable gardening, rabbit keeping, and poultry and vegetable keeping. Project work should essentially be an educational process. In addition to offering much-needed extra cash or food supplies, the projects ought also be instructive and a means of giving young people knowledge and skills that will be helpful in the future. The projects must be successful because failure might easily result in early dissatisfaction. Since the extension agent is essentially preparing and developing for the future, he should thus devote the same careful attention to youth work as he does to more general extension work. Lastly, an extension agent should feel comfortable interacting with rural adolescents and have a general understanding of their goals and perspectives. Therefore, it could be preferable if younger agents in an extension agency assumed responsibility for youth extension activity. If extension work is to be successful, young people will need to be able to relate to the agent and be willing to cooperate and trust him.

5.6.7. Extension and the Landless

Landless agricultural families can be found in many of the regions where extension agents operate. In addition to those who own no land at all, families whose landholding is insufficient even for subsistence farming are also considered landless. To make ends meet, these two kinds of families must sell the labor of their members.

It is not possible to provide statistics on landlessness worldwide in this book. However, there is evidence that landlessness is widespread and on the rise in many regions of the world, which poses a unique set of challenges for extensions. When extension and farmers have been covered in this guide thus far, along with new ideas, technologies, and practices, it has typically been assumed that farmers had access to the resources (land, for example) necessary to benefit from these innovations. However, many farm families around the world lack direct access to the resources needed to implement the agricultural improvements that the extension service recommends. Therefore, it is necessary to consider, "What is Extension's responsibility towards these families"? Extension agents face a great challenge when working with these people, but they can start by learning about the traits of landless families.

1. The landless lack an economic base on which to build any kind of future.

2. They are dependent upon others for their livelihood, under conditions over which they have little control.
3. The landless family's livelihood is precarious.
4. They have little contact with extension or other government services.
5. They have no influence over the decisions that affect their family's livelihood.
6. Few organizations represent their interests.

It is evident that the aforementioned provide a significant obstacle for extension agents who encounter landless households in their service regions. In order to accomplish the goals of rural development, the agent will already need to execute policies and programs, meet targets, prioritize raising the productivity levels of currently available resources, and collaborate with farmers who have the capacity to produce more: However, extension services generally (and agents specifically) ought to include the situation of the landless whenever feasible. The agent's ability to commit time and money to enhancing the landless people's standard of living will be the determining factor. At the very least, the agent ought to invest some time in learning about the issues facing the landless in his community and be on the lookout for opportunities to contribute. Extension services have attempted to address the issues faced by the landless in a number of countries, including Peru, Bangladesh, and Nepal. These initiatives have been grounded in three primary initiatives:

1. **Organization:** The extension agent supports organizations that are formed to speak for the interests of the landless in the same way that he would help a group of farmers. However, considering the aforementioned traits of landless households, the agent needs to be ready to invest a significant amount of time before any organization forms.
2. **Resources:** When resources are available, the agent should work to help the landless people develop their smallholding or, in the event that land becomes available, assist them in gaining access to it. Little stock-raising initiatives and collaborative craft manufacturing ventures with family members will also generate some revenue.
3. **Motivation:** The desire or drive to work towards bettering their situation is frequently absent from landless people. The extension agent can express his desire to assist them, offer his support, and generally try to pique their interest in activities that will better their life.

To say that an extension agent can resolve the core issues facing the landless would be incorrect. The landless people's predicament is rooted in rural society overall, and their standard of living won't improve unless there are improvements made in the rural landscape. A farm family's lack of resources should not be a reason for the extension service or the agent to turn them away; extension is for the benefit of the entire rural population. The agent needs to learn about the structural barriers preventing the landless from developing, and he needs to make every effort to remove these barriers. Above all, he needs to acknowledge the landless people and extend his support in helping them obtain a more stable means of subsistence.

Chapter 6

EXTENSION REFORMS

Contents

6.1. OVERVIEW

In the past, public extension and rural consulting services have been crucial in helping numerous nations increase agricultural output. On the other hand, the effectiveness of improvements in agricultural extension systems has varied. Although a number of actions have been made to improve the effectiveness of the extension systems, it is still difficult to apply the knowledge gained to policy and extension reforms. It makes sense that before seriously considering overhauling and modernizing the current national agricultural extension services, policymakers would first need to be persuaded of the critical role that extension plays in national development. This is significant because, regrettably, there is a generally negative perception of extension. This is due to a number of factors, including a weak lobby for extension in the majority of developing nations, poor organizational setup at first, a general lack of trust in extension among research organizations, and historically subpar career development opportunities in the extension field. Extension services have been a primary target of downsizing since the beginning of structural adjustment programs in developing nations many years ago. Governments have occasionally been led in this way by the relevant donors, but occasionally the value of extension has been questioned by powerful lobbies in specific agricultural fields, most notably agricultural economy and research. It is quite challenging to carry out extension activities in developing nations. It is typically done in the field in inclement weather and with the barest minimum of facilities—not in air-conditioned offices or labs by people dressed in formal attire. It is true that working with largely uneducated and impoverished rural people on an extension mission to positively influence their behavior is a more difficult challenge than working with plants and animals in the comfort of research stations. Indeed, this is unfortunate for the emerging nations. Most agricultural academics have considered extension as a subpar issue, even though there is evidence that poor extension prevents farmers from using certain otherwise outstanding technologies. Extension made a substantial contribution to the Green Revolution in the 1960s by carefully introducing and promoting high-yielding wheat and rice varieties and by using farm inputs in accordance with research recommendations. Research agendas in agriculture remain mostly academic until extension staff offer feedback regarding the farmers' identified and unresolved field difficulties. While extension focusses on how users will accept and use those technologies, research concentrates on the technical aspects of creating viable technology. Strong extension services are essential for applied research institutions to operate in a field problems-oriented mode, and strong agricultural research institutions are necessary to support the extension services in order for them to effectively serve rural communities. It is unfounded for academics to think that if they create a useful technology, farmers will immediately accept it and no extension is necessary. To the dismay of good scientists, hundreds of excellent inventions created by researchers would not be on shelf space, as was noted in a significant international meeting that took place in Addis Ababa in 1995 and included the Sasakawa Foundation.

The good technologies must first make the distance between the farmers' fields and pertinent research institutes in order for the farmers to consider them for potential adoption. Subsequently, they ought to be presented to the farmers using simple language, and the benefits of the technology over conventional methods ought to be persuasively illustrated, for example, by means of a field demonstration. The components that must be included in order to test the new technology, including the cost and any potential risks, must then be described. Afterwards, a variety of adult education techniques and participatory decision-making strategies must be used to promote debate on the data presented on the new technology in order to evaluate both advantages and disadvantages. It will be necessary to make plans for other farmers to take advantage of this brief opportunity for "sample adoption," assuming that a few forward-thinking farmers are open to experimenting with the new technology. It will be necessary to regularly monitor the technology trial and have conversations with the progressive farmers in order to troubleshoot any issues and, eventually, evaluate how well the new technology performs overall in *in situ* field circumstances. It is necessary to inform the relevant researchers who advised the specific technology of any positive or negative outcomes, along with any plausible explanations. The technology will be more likely to be adopted by other farmers if the advantages outweigh the drawbacks; if not, it will need to be removed from the list of recommended extensions as an inappropriate technology. This is an oversimplified description of the process that each extension worker must go through each time researchers offer a new technology. Who will be in charge of adhering to this protocol in the event that there are no extension workers present? The researchers themselves might have the answer. If this response is accepted, the researchers will actually be carrying out the extension function—even if they are not affiliated with any extension service—which is totally permissible. Would this idea, however, actually work in practice, given that the majority of researchers would rather remain inside the walls of their institutions and rarely, if at all, visit farmers' fields? Thus, it stands to reason that if extension services are discontinued, researchers ought to be prepared to receive training and take on the role of extension workers—a situation that is most likely unappealing to them right now. Historically, tiny extension units were established by agricultural research institutes in several developing nations as part of their organizational structure to disseminate their technologies, primarily through publications. Even if the researchers were happy with their papers, the notoriety they received was rarely shared outside of the institutes.

Later, these units were dissolved. Research experiments were carried out in farmers' fields as part of a number of research-outreach projects that were also tried. This strategy had more favorable outcomes, but only in those locations where farmers and extension agents actively participated in the entire process. However, the approach was not a good replacement for a working extension service.

Strong extension services, initially provided by the public sector and currently by the public and/or private sectors, have always been available in highly developed agricultural nations like the United States of America, Canada, Australia, and Denmark. Their farms—which are primarily commercial—their farming operations—which are primarily mechanized—and their farmer population—which is primarily a small portion of the nation's population—may all differ greatly from those of developing nations, as do their extension programs. When allocating resources, the discipline of extension has never been viewed as less important than other agricultural disciplines in any of these highly developed nations. Over the previous 10 years, major donors such as the World Bank have seen a nearly steady decrease in the amount of resources allocated to the agriculture sector; however, in recent times, this trend has clearly reversed. There are a number of reasons for this, but the main one is that wealthy nations are determined to combat poverty and hunger in developing nations as these two threats have resulted in a wide range of grave issues with global implications. The stated United Nations Millennium Development Goals, which are closely related to the extension function, include eradicating extreme poverty and hunger, promoting gender equity and women's empowerment, fighting HIV/AIDS, malaria, and other diseases, and ensuring environmental sustainability. Furthermore, the necessity of sustainable rural and agricultural development is still being emphasized globally. These facts have sparked a global movement to reform and modernize the established extension systems in order to enable them to successfully fulfil the expected role, in addition to reigniting interest in the potential power of extension. The new extension services must possess the following qualities: expertise, resourcefulness, a wide technical mandate that goes beyond technology transfer, and true efficiency regardless of the modality—single, pluralistic, public, private, or a combination of the two. Any money spent on modernizing and reforming extension would undoubtedly benefit the nation in the long run. However, in the event that any developing nation lacks an efficient extension service during this critical juncture in world development, its farmers will almost certainly be severely undereducated, its agricultural sector will likely remain underdeveloped, and rural communities will likely face extreme poverty.

6.1.1. Formal Introduction of Extension in Developing Countries

Agricultural extension (because it applies the function to agriculture) was formally introduced in the 1950s by visiting American professors, technical experts in agricultural and rural sciences, national government officials, and university faculty who received education or training in the United States of America. Although the function of extension has been performed informally in developing countries for centuries, their application to agriculture led to this development. They drew inspiration from the excellently organized agricultural research, extension, and education efforts conducted by the American Cooperative Extension Service and its affiliated Land Grant College system. The primary purpose of the agricultural extension organizations established in emerging nations was to distribute improved technology to farmers by means of agricultural research institutes. Oral guidance was given to individuals or groups of farmers in their homes or on their farms. This was supported by examples of how suggested technology were applied in farmers' fields and the positive outcomes that resulted. Printed materials, radio, and eventually television and video were among the numerous communication channels that helped spread knowledge of technology, smooth its transfer, and possibly even encourage farmers to use it. In most nations, the provision of agricultural inputs—chemical fertilizers, insecticides, and herbicides, in particular—was also handled by extension agents.

6.1.2. Extension Function Vs. Extension Organization

The extension role is crucial for the welfare of farmers, much like many other vital daily functions like health and education. It doesn't matter who does it as long as it's done well. In addition to government extension departments, other potential actors in the extension space include non-governmental organizations, universities, research institutes, farmers' associations, private extension service providers, and private extension consultants. Recent developments may also reveal other players. On the other hand, extension organization refers to the way the department or agency in charge of the extension function sets up shop in order to carry out this duty. This is the distinction between "extension organization" and "extension function." It is a prevalent misconception that criticism of any public agricultural extension organization for its subpar performance is directed towards the extension function, even if this is completely untrue. It seems to reason that the extension function cannot be held accountable for the extension organization's subpar performance. The public extension function is becoming more and more important, especially in light of the increased global emphasis on ending hunger and poverty. If the public extension services are not meeting these goals, they should be strengthened, reformed, or eliminated entirely. In any case, alternative delivery methods should be explored. The real question is how the millions of farmers in that nation will receive education in the areas of personal, professional, and village development if the extension function also suffers during the dissolution of an inadequate extension organization by any government due to financial constraints and occasionally because of pressure from some donors. When making judgements on investments in extensions, the policymakers of that nation should always put this question in the back of their minds.

6.1.3. Increasing Diversity of Extension Function Players

Agricultural extension services were formally under the Ministry of Agriculture from the outset, when extension was first established in emerging nations. As a result, the government was left to handle the extension function on its own. As time went on, departments within the Ministry of Agriculture or other ministries with a focus on agricultural and rural development were frequently enticed to start their own extension services with whatever staff they could muster up in the field, no matter how little. Thus, in addition to the primary agricultural extension service, other services including livestock extension, plant protection extension, marketing extension, and horticulture extension arose. This trend placed a great deal of government extension workers in the field, many of whom were loyal to different technical departments but who suffered from low pay, poor career opportunities, poor career funding, and most importantly, a lack of coordination between different extension units. Farmers had to put in a lot of work to comply with this structure, which occasionally even led to misunderstanding due to conflicting or redundant technical advice. But since farmers received this kind of free extension advice, there wasn't much accountability for the extension agents. The establishment of semi-governmental autonomous authority for certain agricultural commodities in numerous emerging nations is a noteworthy development. Jamaica, for instance, established several such organizations under the names Coffee Board, Sugar Board, Tea Board, etc. These powerful and wealthy organizations established their own mini-extension services by luring qualified extension workers and subject-matter specialists from the government extension service with contracts that offered better pay, benefits, and vehicles than those offered by the government service, since the goal of creating such bodies was to ensure adequate and proper production and export of these valuable cash crops for earning foreign exchange. Undoubtedly, this approach gave commodities production and export companies access to the best extension specialists available, but it also took these skilled and knowledgeable employees away from the public extension agencies. Additionally, small farmers were not receiving good extension guidance from these extension professionals because their primary concentration was on commercial farmers. Sales representatives from private enterprises that dealt in chemical fertilizers, pesticides, insecticides, and herbicides began giving farmers extension guidance while government agencies were vying for more field staff. Their primary goal was to increase product sales, and, if any cooperation or consultation at all, there was very little with the government agencies. Many of these commercial companies have operated and maintained their own mini-extension services since the 1960s. Recent years have seen the rise of private and civil society actors who are interested in providing extension services to farmers at no cost or for a fee, as a result of mounting economic pressure on governments and criticism of the steadily deteriorating performance of public extension services. Pluralistic extension delivery patterns are still in their early stages of development, and in numerous countries, both advantages and disadvantages of the modality have been shown. In light of the knowledge gained thus far, a process of making beneficial modifications has already commenced and will undoubtedly last for many years. Public extension services have either been wholly or partially privatized in a few affluent nations.

6.1.4. Challenges to Traditional Extension Practices

Global focus on sustainable development, encompassing rural development and agricultural progress, along with trends like globalization, liberalization of markets, decentralization, privatization, and democratization, are generating new educational needs for developing country farmers, whether they are subsistence or commercial. These demands are posing a challenge to long-standing regulations and procedures inside conventional extension systems, particularly in light of the information technology revolution. In order to enable the extension function to be carried out with excellence in line with the global challenges to their economies, and particularly to their agriculture sector, policymakers in developing countries would be well advised to examine and challenge the discipline of extension within a global context. It will not be very beneficial to make cosmetic adjustments to the current national extension systems or to repeatedly train people in stereotypical agricultural topics. Beat a dead horse just as well.

6.2. DEVELOPMENTS DRIVING EXTENSION REFORMS

Like the majority of other disciplines and institutions, agricultural extension organizations are not immune to the numerous advances occurring in their surroundings. In this part, a few significant advancements and their effects on several facets of extension are discussed.

6.2.1. Globalization and Market Liberalization

The goal of globalization is to increase trade, information exchange, and financial ties between nations while promoting open competition by removing trade and other national safeguard barriers. This is supposed to provide producers of goods and services, regardless of where they reside, with fair and relatively equal conditions, thereby enabling them to benefit from international contacts and opportunities. Regardless of the benefits, drawbacks, opportunities, and dangers of globalization, the contentious issue is the lack of certainties regarding just and equitable working conditions for all producers. In the current global market, developing countries are unable to compete with developed countries in terms of production, quality, and exports due to a variety of issues, including a lack of resources, a lack of technology, weak rural institutions, inadequate infrastructure and communication systems, etc. The dominant theme conveyed by the well-known protests at economic summits is that globalization will increase the wealth of rich countries and decrease the wealth of disadvantaged countries. Rich and poor countries seriously disagreed on agricultural matters at the September 2003 WTO summit in Cancun, Mexico, which resulted in a sad failure. Beyond politics, farming communities in less developed nations will undoubtedly face hazards as well as opportunities as a result of the globalization phenomena. As a result, the communities need to be informed and ready to adjust their farming practices in light of globalization. This is a duty that will mostly fall on the national agricultural extension systems, and they need to start preparing early to meet the task. Globalization cannot exist without the liberalization of markets. This demands the lifting of trade restrictions, tariffs, and other laws that are typically imposed by nations to safeguard their own businesses and goods and to deter the import of goods made elsewhere in the world. Liberalization calls for deregulation, or the opening of markets, to allow for the free flow of products between nations. In order for the market to fairly determine the price of different goods and commodities based on supply, demand, and quality and for consumers to freely choose what is best for them, it also advocates for the elimination of artificial price controls and public support for the farming sector in the form of farm subsidies. However, large government subsidies in wealthy nations already provide their farmers a significant advantage over small and impoverished rival farmers in less developed nations, endangering their very survival. For instance, if cotton—which is produced in wealthy nations at a low cost due to government subsidies—continues to be dumped and sold cheaply in the Malian market, small farmers in Mali who have relied on cotton cultivation for centuries as their primary source of income could be completely destroyed. Market liberalization will inevitably have an impact on developing country farmers, whether they farm for profit or for subsistence. As such, national agricultural extension services should have the expertise to advise farmers on how to effectively navigate a liberalized market. Millions of farmers who cultivate only enough for sustenance sometimes generate a small amount of extra harvest to sell at the market during exceptionally good cropping seasons. Today, nevertheless, there is a determined effort to commercialize farming, even on a small scale. Asking the very poor farmers in remote, marginal areas to migrate elsewhere because it is not cost-effective to invest in their development raises both ethical and technical questions about whether it makes sense to let them continue as subsistence farmers have for generations, or to allow their operations to be transformed into agri-business, rural enterprise, rural industries, or other commercial ventures. These are reasonable issues given the evidence that, in many countries, rural poverty has not improved, if anything, and that, in contrast to their parents, young people from rural areas are less likely to remain in their villages and carry on farming. If privately owned and farmed small land units are to be combined or merged to produce larger, commercially viable plots, appropriate land consolidation legislation and agricultural policies will be needed. In order to enable subsistence farmers profitably commercialize their businesses without sacrificing their sense of pride in owning their land, agricultural extension services will need to devise solutions. The topic of creating and consuming genetically modified organisms (GMOs) is a hot topic of discussion worldwide. Different points of view are represented by the various stances. While it is important to maintain biodiversity and raise public awareness of the potential health concerns associated with genetically modified food for humans and livestock, there is also a need to feed growing populations using all available means. The farmers who grow food for millions of people in underdeveloped nations are the least knowledgeable parties in this argument. To date, no agricultural extension service has felt compelled to inform farmers about this contentious but crucial topic. In underdeveloped nations, biotechnology is still relatively new, thus there is no pressing need to launch any special extension initiatives. However, now is the exact moment to start putting together training programs and instructional materials for farmers and extension agents alike.

6.2.2. Privatization

Particularly in industrialized nations, the private sector now oversees and provides many services that were previously overseen by governments. Several developing nations are doing the same. The fundamental cause is the public institutions' declining finances, which render them comparatively ineffective and less productive. This results in financial losses for the government as well as unhappiness among the populace. In contrast, the private sector is more driven to provide its customers with effective and superior services since it typically possesses greater resources, creative ideas, and a profit-making incentive. A number of bilateral and multilateral funders have promoted the privatization of national extension services in recent years. Numerous arguments are made in favour of privatization, such as the following: public extension systems have failed or are at best ineffective; farmers benefit from extension advice, and government budgets for public services are shrinking; the private sector actively sells farm inputs and machinery; therefore, it makes sense for it to handle providing advice to farmers on agricultural matters; healthy competition among service providers will result in better quality and lower costs for services; it is prudent to consider financing and delivery of extension services as separate functions, to be performed by different sectors, that is the customarily cordial and informal connection between government extension workers and subsistence farmers in developing nations will undoubtedly be upended by the trend towards privatizing extension services, since farmers have never before been required to pay for extended advice. Beyond the question of whether national extension services and their clients need to be well-versed on the advantages and disadvantages of this significant topic, there is also the matter of whether whole, partial, or no privatization of extension is necessary in developing countries. In a number of ways, a number of affluent nations have privatized all or part of its agricultural extension services. Concepts like as contracting out, cost-recovery, and outsourcing are associated with the privatization movement. With a particular system in place in Costa Rica, farmers receive vouchers from the government that they can use to receive extended help from private experts. Over time, the public extension service in England changed to become a private consultancy firm. Improvements in staff productivity are the positive outcome; nevertheless, small farmers who are unable or unwilling to pay will lose out on extension services. There are also rumors that the government has taken over several advisory roles that were previously privatized due to discontent with the business community. In the Netherlands, farmers contribute roughly 60% of the extension budget, with the government covering the other 40%. Enhanced productivity, better quality, customer focus, employee work happiness, and more farmer marketing opportunities are among the advantages. Loss of government authority, the government's failure to fulfil financial commitments, and weakened stakeholder communication as a result of increased competition are among the issues. The private sector's entrepreneurial efforts to establish a lasting partnership with farmers in Albania have shown to be effective. In Nicaragua, the extension services are semi-private and decentralized. To create a private extension service, Bulgaria privatized certain state farms to be utilized as demonstration farms. The government chose to create the National Agricultural Advisory Service (NAADS) with outside funding support because the experiment was a failure. With support from the FAO, the NAADS is currently being assessed for potential revision to better serve farmers' evolving learning needs, given that the nation will likely join the EU in 2007 or shortly thereafter. In Estonia, there is a private extension advisory service for wealthy farmers as well as a public service for less fortunate farmers. Through the establishment of a pool of private extension specialists within its current public extension service, Uganda is experimenting with the privatization of extension. Registered farmers' associations will be able to bid on services related to specific enterprises from this pool, and will be able to pay for those services with funds provided by donors through decentralized government units. Although this arrangement is a risky experiment, it is unclear how long it will last if the donors' money runs out. Attempts to partially or completely privatize national extension services in Israel have failed. Although the government still provides extension advice, it encourages privatization by allowing growers to donate a portion of their income to research and development, which includes extension. Public and private partnerships finance and run extension service units, and growers pay for services beyond the basic package provided by commodity production and marketing boards. Special agreements with commodity farmers' organizations allow extension staff to work on their days off in exchange for direct payments from farmers. Growers associations provide mobile phones to extension advisers, and farmers directly pay for their participation in training programs. Farmers ought to foot the bill for the advice provided by extension agents, according to proponents of privatizing extension services. Nonetheless, there is legitimate concern that small farmers might lose out on the services due to the fervor for privatization. The small farmers are either unable to pay for the extension guidance, or they do not think it is worthwhile. Conventional wisdom dictates that in developing nations, extension counsel should be financed by commercial farmers and large cooperatives, while small producers should continue to receive free extension services from the government. Additionally, steps must be taken to shield farmers from being taken advantage of by the private industry.

6.2.3. Pluralism

The government now provides several public service sectors to private businesses as a result of the growth of privatization. In industrialized nations, government monopolies are not the most preferred choice, particularly when it comes to development when there are capable non-public institutions in place.

Using many organizations—public or private—to provide farming communities with extension services is becoming a more frequent approach. The obvious justification is to combine all available resources to relieve the burden of tight budgets and understaffed ministries of agriculture while also allowing farmers to take advantage of several sources of income. Pluralism is already being practiced in many emerging nations. In Zimbabwe, farmers receive extension guidance from a large number of non-governmental organizations, for-profit businesses, and semi-autonomous organizations. For the first time, under a decentralized extension system, Bangladesh's national policy on agricultural extension acknowledges the necessity of collaborations with NGOs and the corporate sector. Approximately 50,000 farmers, primarily in remote locations, are reached by 70 NGOs in Honduras, where small farmers are unable to pay for the privatization of extension services. Certain initiatives in Indonesia have promoted the participation of agricultural research institutes, agricultural universities, farmers' associations, and the business sector in the provision of extension services, in addition to NGOs. Installing a functional pluralistic agricultural extension mechanism is primarily challenging because it requires excellent coordination across organizations; otherwise, there may be competing technical recommendations, which could confuse farmers. Governments ought to be in charge of organizing, overseeing technical projects, and maintaining product quality.

6.2.4. Decentralization and Devolution

Numerous nations have embraced decentralization and devolution, and this tendency is still going strong. The primary justification for the change is that decentralized administration transfers decision-making power to lower administrative and political level entities, who are purportedly better equipped to organize and carry out development projects with participants' active participation because they are physically closer to the people. Although the idea is sound, there are a number of reasons why its application in developing nations has not gone well thus far. For instance, some central governments find it difficult to give up control over decision-making, particularly when it comes to financial matters; decentralized government units struggle to manage affairs; and local politicians unnecessarily meddle in technical areas in which they lack the necessary expertise impedes implementation. Decentralization has frequently resulted in the ministry of local government handling agricultural extension duties. Decentralization is a fine idea in theory, but there are significant obstacles to its further development at this point. For instance, in the Philippines, the meddling of politicians at the municipality level has hampered the efficacy of extension programs to the point where candidates without agricultural degrees have been hired for open positions in agricultural extension. Additionally, there has been a negative impact on extension staff members' professional advancement chances. According to current thought, the province should once again assume responsibility for extension services since municipalities are not sustainable economic entities. Extension services have been marginalized in Indonesia because, in spite of national policy, the majority of district governments prioritize projects that could yield immediate cash, and agricultural extension—which, unlike estate crops or livestock, typically yields a long-term return—is not one of those projects. Field-level decentralized extension services in the United Republic of Tanzania are plagued by a shortage of operating funds. The decentralized government units in Kosovo, where UNMIK (United Nations Interim Administration in Kosovo) expatriates make up the government, are reluctant to work closely with the central administration. Some district authorities in Uganda have chosen to build feeder roads using the extension budget, delaying the pay of extension employees for several months. The lack of proper preparation by the central government or the local governments prior to decentralization is one factor contributing to the "institutional chaos" that many nations undergoing decentralization and devolution are experiencing. Decentralization is a real two-edged sword that, in the wrong hands, may do more harm than good. If extension services are not adequately prepared for this crucial administrative, technological, and physical shift, they may experience issues with marginalization and a loss of professional identity as more and more nations adopt decentralization policies.

6.2.5. Client Participation in Decision-Making

Civil society is increasingly advocating for the reasons that democracy, inclusive decision-making, openness in public affairs, and good governance are essential to ending hunger and poverty, eliminating corruption, distributing benefits fairly among the different segments of society, helping the weaker and more vulnerable members of society, making the best use of human and material resources, establishing sustainable livelihoods, and advancing human development as a whole. In many cases, even the financial aid provided to developing nations by bilateral donors has been contingent upon the government's track record of upholding human rights and promoting good governance. In recent years, many tactics have developed that purport to guarantee inclusive decision-making and the participation of all parties in collaborative planning and execution. These conceptual thrusts have affected all service institutions operating at the grassroots level. A number of extension emphases, including working through farmers' groups, preparing, and delivering client-oriented messages, adhering to gender sensitivity, and establishing links between research, extension, and farmers, have been brought about by the strong trend towards empowering farmers through active participation in decision-making. A few other participatory instruments have also been developed, such as the KAP (knowledge, attitude, and practice) survey and the PRA (participatory rural appraisal). Farmers, researchers, and extension experts have come together to form the Agricultural Technology Assessment Institutes that Indonesia has successfully developed at the provincial level. Under the FAO Special Program on Food Security, groups of highly motivated small farmers were established in Pakistan. These groups have been making decisions about marketing produce, diversifying income sources, cultural practices, farm machinery, quality seed, fertilizer, water management, and group cash savings. The government has started to replicate the strategy in thousands of communities because the yield increases from the pilot project have been so good. The FAO has implemented a demand-driven, gender-sensitive, bottom-up extension program in the northern region of Pakistan, wherein groups of men and women villagers are involved in creating grassroots plans for services that they would like to receive. In Argentina, the participation of all stakeholders in significant decision-making is a key component of the success of a federal program for small and medium-sized farms. Comparably, the United Republic of Tanzania created Participatory Farmer Groups comprising both male and female farmers, while Jordan's extension department established Close Contact Groups of farmers. Extension workers, farmers, and other stakeholders are involved in the preparation of an extension delivery partnership mechanism in the Philippines, where extension services have been devolved. The citizens have been helped by an FAO project to outline a methodology for participatory, grassroots extension program development.

6.2.6. Natural and Man-Made Disasters

Unfortunately, in this day of advanced science and civilization, human suffering is increasing due to both natural and man-made tragedies. A great deal of physical and human suffering is caused by battles, wars, famines, droughts, earthquakes, hurricanes, and epidemics, all of which disturb the usual stream of existence. The number of nations experiencing human-caused disasters has sharply increased, rising from an average of five in the 1980s to 22 in 2000, primarily as a result of armed wars. By the end of the twentieth century, thousands of millions of people were negatively impacted by conflict in 30 to 40 different countries. Land mines in some places prevent farmers in nations like Afghanistan from farming. Soil, water, and seeds—basic gardening supplies—are not easily accessible. Farmers no longer have access to expert assistance because extension services have vanished. The damage to the infrastructure prevents farm inputs from being delivered or transported. Food distributions are effective only to a certain extent, despite the efforts of NGOs and food relief agencies in disaster-affected areas. The number of weather-related and other natural disaster situations annually has increased from 10 to 18 in recent years. Approximately 32,000 people lost their lives and 300 million were uprooted from their homes and means of subsistence in 1998 alone. Thousands of people were killed by the December 2004 tsunami in several nations, and the farming infrastructure along the coast was completely destroyed or severely damaged. Over 50,000 people were murdered in October 2005 by a powerful earthquake that predominantly impacted the Pakistan-administered Azad Jammu and Kashmir territory. The victims were mostly small-scale farmers from mountainous regions, and the earthquake also severely damaged physical infrastructure and means of subsistence. Though it's true that agricultural extension services haven't yet had a major impact on the recovery of agricultural operations following disasters, they are being trained to help in post-disaster situations in some countries, such as Bangladesh and the Caribbean, where powerful storms occur annually. Extension services should collaborate with relevant institutions to help the rural population grow food for survival using whatever resources are available, as it is recognized that they cannot meet this enormous challenge on their own. We're still waiting on a thorough answer from national agricultural extension systems.

6.2.7. Information Technology Revolution

The world has almost shrunk and is being impacted by the information technology revolution that we are currently experiencing in almost every aspect of life. Extension services shouldn't be reluctant to make use of information technology since it is a fantastic tool that may be used by organizations to advance humankind. They can take advantage of this opportunity to develop their own skills and educate the rural masses that have access to the media. When it comes to meeting with farmers and researchers in person, the two main obstacles facing extension organizations in developing nations are physical distances and a lack of transportation options. By creating and utilizing suitable, interactive information methods, the new information technology could largely get beyond these physical limitations. Advances in agricultural and rural development are already being made possible by advanced information technology. Telecentres are being tested in numerous developing nations, including Mali, the Lao People's Democratic Republic, and Vietnam. These facilities have previously shown promise in a number of Western European nations. Research and extension are becoming more closely linked through the creation of virtual links. One such example is the VERCON (virtual extension, research, and communication network) tool that the FAO has implemented in

Bhutan and Egypt. Internet access and interactive email services have been set up at the municipality level in the Philippines as part of an FAO project to help decentralized extension workers. Additionally, expert systems are being created to somewhat offset the all-too-rare visits by subject-matter experts to farmers' fields. In Bangladesh, cell phone usage has become commonplace, and the devices are utilized for rural development initiatives. In Estonia, more than 30% of extension workers utilized the Internet. On the Internet, one can obtain applications such as "virtual farms" and "virtual gardens." The primary concern is how to use cutting-edge information technology to the advantage of farmers and extension agents while maintaining the value of distinctive local elements like traditional communication styles and without thinking of information technology as a substitute for extension agents, who are still highly valued human resources.

6.2.8. Rural Poverty, Food Insecurity, and HIV/AIDS Epidemic

Internationally, there have been repeated calls for the eradication of hunger and poverty in rural areas, as well as for support for marginalized populations including women, youth, and the rural landless. The most recent of these calls is found in the Millennium Development Goals. The percentage of the impoverished has increased recently. Furthermore, despite the fact that there is enough food produced for everyone on the planet, over 800 million people, or about one in seven, do not have enough to eat. Vulnerability, poverty, and hunger are related. While the world celebrated the start of a new millennium, 13 million people in the Horn of Africa were on the verge of starvation, and foot-and-mouth and mad cow diseases crippled European farmers and the livestock industry. This was followed by an outbreak of avian flu in several Asian nations that has now spread to Europe. In 2005, a massive famine devastated Niger, killing thousands of people, many of them children, as the G-8 leaders pledged billions of dollars for the development of Africa. Estimates from the UNAIDS/WHO AIDS Epidemic Update (December 2004) state that by the end of 2004, 37.2 million adults and 2.2 million children had HIV; 3.1 million of them had died from the disease. Approximately 95% of all HIV-positive individuals worldwide reside in underdeveloped nations. The AIDS epidemic has severely impacted Sub-Saharan Africa, making it the most afflicted region globally. Although barely 10% of the world's population resides in the region, more than 60% of all HIV-positive individuals do. In 2004 alone, an estimated 3.1 million adults and children contracted HIV. In Asia, an estimated 1.2 million people contracted HIV in 2004, increasing the total projected number of HIV-positive individuals to 8.2 million. Concerning statistics have been documented for more global regions. Unquestionably, there are detrimental consequences on the labor market that cause workers in all fields to lose their training, expertise, and experience. Farm labor, which was formerly in great supply, is quickly running out. Research from the FAO in Malawi, Uganda, and Zambia, among other studies, demonstrates how the epidemic has impacted agricultural extension organizations by making outdated, traditional extension methods ineffective in addition to causing staff deaths and extended absences. This is because labor-intensive cropping patterns have changed due to factors like farmers' deteriorating physical health, a shortage of farm labor, the emergence of "new farmers," who are primarily elderly, widowed, and have small children, outdated farm tools, and outdated rural credit approval standards. Extension agencies in HIV/AIDS-affected nations, particularly in sub-Saharan Africa, need to develop new extension tactics that are backed by adjusted farming practices and useful agricultural equipment. Since the World Food Summit in Rome in 1996, there have been numerous new anti-hunger efforts including all the institutions that work with rural populations. NEPAD, or the New Partnership for Africa's Development, has also been introduced. Agricultural extension agencies that frequently interact with rural populations would find it challenging to stay apart from these global initiatives combating HIV/AIDS, food insecurity, and rural poverty. To effectively contribute to the fight against these human miseries, extension services must build the necessary capacity as well as appropriate extension educational methodologies and instruments. The FAO has created helpful resources that field extension agents can utilize to teach farmers how to reduce the spread of HIV infection.

6.2.9. Integrated, Multi-Disciplinary, Holistic, and Sustainable Development

It is increasingly undeniable that multidisciplinary, integrative, and comprehensive approaches to development are preferred globally. The argument is that multisector development occurring concurrently is more significant than sector development occurring separately. An increasing number of organizations are making structural revisions in support of this decision. The "integrated rural development approach," which was used in several nations throughout the 1960s, is making a comeback in a modified version that incorporates gender-sensitive methods, stakeholder participation, bottom-up planning, and a focus on ending rural poverty. Many multilateral donors have moved away from the previous project model and towards a program strategy. In Namibia, Uganda, Zambia, and Zimbabwe, FAO has implemented an integrated program that includes up to six technical components, including extension and research. Holistic development is also supported by the sustainable livelihoods idea.

The function of extension in a broader framework of rural and agricultural development has been investigated by the Neuchatel Initiative Group, which is made up of significant funders interested in changing national agricultural extension systems. The Agricultural Knowledge and Information Systems for Rural Development (AKIS/RD) project, which aims to increase collaboration and cooperative operations between research, extension, education, and farmers, is a joint effort between FAO and the World Bank. FAO has carried out research on AKIS/RD in 10 different countries throughout the world. These studies were then analyzed and synthesized in order to draw conclusions and offer suggestions for AKIS/RD actors to get in closer contact with one another. Various agricultural line departments are still tempted to maintain separate extension services, but due to well-documented and observed issues with this approach, extension services are being unified in a multidisciplinary manner to maximize resource utilization and streamline bureaucracy. It is true that the time of farmers cannot and ought not to be wasted on the one-on-one visits from numerous extension agents, each of whom specializes in a different area of agriculture. It is common practice to establish or fortify teams of multidisciplinary subject matter experts throughout the decentralization of extension services in several nations. FAO has given Uganda technical support to incorporate HIV/AIDS education and agro-forestry into agricultural extension initiatives, resulting in a multidisciplinary extension strategy.

While working with the Iranian government to restructure the national extension system, the Food and Agriculture Organization (FAO) has suggested that multidisciplinary teams of subject-matter experts be assigned to districts, with the makeup of each team depending on the district's priorities and technical requirements. As a result, there will be a fisheries specialist in coastal regions and a livestock specialist in districts with a predominance of animal operations.

In Indonesia, a project funded by the World Bank has combined the extension services for forestry and agriculture into a single program. In the Democratic Republic of Laos, the same thing has been done. According to the Agriculture and Fisheries Modernization Act, extension services are provided for both agriculture and fisheries in the Philippines.

The unification of the current extension services is one of the reform measures being implemented in Vietnam as part of a project supported by the Asian Development Bank. Extension messaging should incorporate certain non-agricultural but very relevant components in order to support sustainable development. Egypt has been receiving assistance from FAO to incorporate lessons about environment and population education into its continuing agricultural extension activities. In a similar vein, Pakistan has received support in setting up extension services to facilitate and actively participate in the education of rural populations on sustainable resource management. For many years, the FAO's integrated pest management program (IPM) has discouraged the overuse of chemical pesticides by encouraging biological control of insect pests.

6.3. GUIDELINES FOR REFORMING NATIONAL EXTENSION SYSTEMS

The national public extension systems have come under growing fire in recent years for their perceived lack of effectiveness given the money spent on staffing and organizational upkeep. They have been criticized for being a drain on public resources and, as such, have been singled out for either complete elimination or significant changes. There have been reports of certain extension services being so big that their own weight prevents them from moving around very well. When the financial, logistical, and career development constraints that extension workers are forced to work under are objectively examined, as is the case when these services are compared to other disciplines like agricultural research that have far better service conditions (including an operational budget), some of this criticism is legitimate and some is not. However, it is still the case that, as previously discussed in this chapter, modernization, and reforms are required in the current national extension systems due to the various global forces that are causing the socioeconomic and political landscape to change globally, posing new challenges and necessitating new learning requirements for farmers in developing nations. The recommendations in this chapter could aid in the modernization process if, upon evaluating the state of the national extension system in comparison to the framework provided in the previous chapter, it is determined that the current system is, in fact, mostly out of date. To put it briefly, the goal of modernization is to use information technology tools to the greatest extent possible in order to make the national extension system demand-driven, gender-sensitive, participatory, bottom-up, and a relatively lean organization that could effectively respond to farmers' extension and training needs emerging as a result of globalization, market liberalization, decentralization, and democratization.

Guideline 1

Assess the existing extension organization against farmers' needs and determine whether to strengthen or restructure it:

Every organization is designed and staffed with people to fulfil a certain function within the parameters of a mission and policy. Because they are financed by public monies, public institutions are meant to assist the general population. Regarding agricultural extension, the organization's purpose is to meet the requirements of both male and female farmers in terms of extension, education, and training. An organization should be examined if it is determined that it is not accomplishing its goals effectively and efficiently. Assessing it based on the farming communities' needs for training and extension will be a quick approach to do it.

Key Actions

1. Assessment of extension and training needs of farmers:
   - i. Preparation of a survey tool, such as detailed structured questionnaire
   - ii. Preparing data-collectors through brief training
   - iii. Selection of survey sample areas
   - iv. Selection of men and women farmer categories sample comprising subsistence, commercial, special interest, landless, etc.
   - v. Survey for data collection on farmers' responses to the questions through individual and/or group interviews
   - vi. Analysis of data, identification of extension and training needs of different categories of farmers, and their prioritization in light of government policies and goals of rural and agricultural development
2. In light of the farmers' top-priority demands for extension and training, the following have been assessed in order to pinpoint specific areas within the current arrangements that require revision, strengthening, or reorganization in order to adequately satisfy the needs of the farmers:
   - i. Existing organizational structure.
   - ii. Operational linkages with relevant institutions.
   - iii. Present mission and technical mandate (whether too narrow or broad enough).
   - iv. Human resources (total number, number of female staff vs. male staff, specializations, qualifications, experience, and location of professional and general services staff).
   - v. Physical facilities (such as offices, equipment, stationery, audio-visual aids, communication facilities, transport, etc.).
   - vi. Operational budget (realistically adequate or not).
   - vii. The existence or lack of non-public organizations that are either highly qualified or reasonably qualified, such as NGOs, farmers' groups, rural community organizations, and private businesses with

sufficient personnel, who are willing to provide extension services to the government under contract.

3. Determine whether it will be more cost- and result-effective by taking into account the gaps found as a result of the assessment of the aforementioned components and the findings of the farmers' extension needs survey:

   i. To continue the public extension organization as it is.

   ii. Or to modify it as necessitated by the field survey results.

   iii. Or to create a mixture of the existing organization and non-public institutions.

4. The decision made on the previous point will guide the creation of an action plan, budget, and timeline for any general organizational modifications, as well as any strengthening or restructuring of particular organizational components, including rewriting of terms of reference, with the involvement of key personnel. Assuming sufficient funding is available, if there is a clear disparity between the technical specialists in the current organization and the technical specializations required based on the needs identified by the farmers, "outsourcing," or contracting out relevant extension work to the non-public sector, should be taken into consideration as an alternative to hiring new technical staff. Time, money, and effort will be saved using this tactic.

5. Implementation of the plan.

Guideline 2

Decentralize extension but not before capacity-building of the staff and orientation of relevant elected officials

Decentralization has proven to be terrible for agricultural extension in some countries, despite being a positive start in the right direction. The temptation for each autonomous district or municipality to follow its own path without considering national policy and priorities is a serious problem, but the unjustified influence and meddling of local politicians, as well as the mindset of local legislators and decision-makers, who fail to recognize the significance of extension in the development of rural and agricultural areas, have done the most harm to extension. This has led to the funds for extension being diverted to other projects, staff members being assigned to non-extension duties, the organization's limited, dilapidated cars and equipment being used for personal use, and the recruiting of non-agricultural graduates for positions in agricultural extension. Due to their heavy reliance on the federal government for financial support, local decision-makers are driven by a strong desire to generate income as quickly as possible in order to lessen this dependency. Agricultural extension, by its very nature educational, is a long process because its goal is to change human behavior, which rarely yields immediate and noticeable results. The local elected officials, whose terms are two to three years, would rather choose to build roads and shopping centers, which would garner support from the public and ensure votes for the next election, or to engage in activities such as raising cattle and estate crops, which yield substantial, palpable, and relatively quick returns. The majority of local decision-makers have low levels of education, thus it is necessary to modify this tendency by educating and sensitizing them to the significance and potential benefits of extension, as well as by orienting them towards national strategy and priorities. In the absence of this crucial group being "converted," agricultural extension under decentralized administration may continue to be marginalized or even cease to exist. Recently decentralizing nations like Nepal and Pakistan can facilitate the process by implementing measures to strengthen the capacity of their decentralized units. These nations can also learn from the experiences of other nations that have implemented decentralization in the past, such as Chile, Colombia, Uganda, Philippines, Indonesia, United Republic of Tanzania, and Ghana. This will assist in avoiding the decentralization-related issues that the "senior" countries encountered.

Key actions

1. Prepare extension staff in advance of the decentralization through the following measures:

- Provide the staff with revised terms of reference.
- Inform the staff about the new administrative and organizational structures. Include information about the Ministry of Agriculture's and local government officials' relationships, the budget allocation process, and any significant effects on staff employment conditions that may arise from the shift to decentralization. Other topics to cover include the staff's physical relocation to new offices at the district or lower levels and the institutional links at the decentralized level.
- Train on the following technical subjects: demand-driven extension, participatory extension approach, grassroots, bottom-up program planning, village extension plans integrated into district development plans, participatory socio-economic impact assessment of extension activities, and decentralization philosophy.

2. Organize specific orientation seminars on extension, or incorporate an extension module into the current training curriculum for local government elected officials. The direction should emphasize the value of extension—which takes longer than short-term investments in cash crops that yield immediate returns—in influencing farmers' positive behavior for sustained rural and agricultural development. Other topics that might be covered include the significance of hiring highly qualified people for extension positions, the necessity of ensuring that the extension staff's terms of reference clearly state that they are not accountable for non-extension tasks, the availability of transport facilities to enable the staff to perform effectively, the importance of the extension workers' focus on small farmers rather than big farmers, the significance of gender consideration in the provision of.
3. It is advisable to think about decentralizing agricultural research concurrently with the decentralization of agricultural extension. This is because it has been noted that institutional connection issues were more noticeable in nations where research remained centralized and only extension was decentralized.
4. Exercise "de-concentration" by moving the majority of the staff to district and lower levels if there hasn't been any formal decentralization but the majority of technical subject matter experts and extension staff are based at the national and provincial levels. This will allow the staff to work closely with the farmers to plan and carry out extension activities. The subject-matter experts should relocate in accordance with the district's technological requirements. For instance, a region that is close to a river or body of water will require a specialist in fisheries, and a district that is mostly home to livestock operations will require a livestock specialist.

Guideline 3

Broaden the technical mandate of extension to aim at broader development of rural human resources

Extension is essentially informal education directed at rural adults who are not enrolled in regular school systems. Its goal is to assist them become more knowledgeable and skilled so they can live better lives. Agricultural extension focusses on providing non-formal education to rural adults, especially farmers, in order to improve their agricultural knowledge and skills and increase farm productivity. This is intended to increase farmers' income and improve their quality of life. Sustainable rural and agricultural development, however, encompasses much more than just using better agricultural technology to increase productivity. It involves managing natural resources, protecting the environment, maintaining a reasonable population growth rate, developing rural human resources, and growing crops, livestock, and fisheries. The Ministry of Agriculture has a higher number of field employees than any other ministry in the majority of developing nations (extension workers). But the current technical mandate of this invaluable field crew is rather limited, covering only the dissemination of better farming technologies to farmers. Because agricultural research institutes do not produce improved technologies frequently enough for extension workers to teach farmers new techniques on a daily or weekly basis, this limited mission results in underutilization of this important pool of field workers. It's also true that government employees don't interact with farmers and communities as frequently as extension agents do. Therefore, having this field workforce offers a great chance to teach rural communities informally the knowledge and skills that could advance the idea and methods of sustainable rural and agricultural development. However, to achieve this, extension personnel's technical mandate would need to be expanded, and staff members will need to get technical and conceptual training on sustainable rural and agricultural development. In order to improve rural areas, it might also be necessary to make some adjustments to the way different technical departments operate.

Key actions

1. Quick training of current extension workers to enable them to educate farmers informally about the principles and methods of sustainable rural and agricultural development through in-service training. The subsequent actions are required:

- Creation of short-term training courses and related materials on the following subjects: the relationship between the environment, population, and agricultural output; the notion of sustainable rural and agricultural development; wise management of natural resources; the impact of any major epidemic, such as HIV/AIDS, on farmers and farming practices and how to mitigate it; how to farm under market liberalization conditions; problem-solving and decision-making; managing farm as business; farmers' group dynamics for lobbying; rural leadership for development; participatory decision-making; gender sensitivity in the planning and delivery of extension services; rural youth development for healthy future; self-help initiative and use of revolving fund for development activities; basics of information technology; food security and how to ensure it throughout the year; off-farm income generation activities during slack period amidst cropping seasons; socio-economic impact of extension activities; grassroots, bottom-up planning of extension programs; post-disaster rehabilitation of agriculture; rural poverty; food security.

- Creating a training plan and carrying it out in order to provide extension workers with in-service training on as many topics as is practically possible. The plan will be based on the established training modules and resources.

2. Expansion of the agricultural extension teaching program at agricultural colleges and universities through changes to the curricula, literature, resources, and instructional strategies.
3. Extension services provide men and women farmers with short-term training in areas such as gender sensitivity, group dynamics, leadership, decision-making, accounting, problem-solving, good governance, citizenship, nutrition, self-help, and initiative, program planning, monitoring, and evaluation, information technology, the value of education for children, particularly girls, networking with other village organizations and farmer associations, etc.
4. In order to coordinate fieldwork for farmers' benefit, the Ministry of Agriculture will establish operational links with other ministries and line departments that oversee technical aspects such as population, environment, and HIV/AIDS that were previously outside the purview of extension workers.
5. Agricultural extension could be renamed as agricultural and rural extension, or as extension for agricultural and rural development.

Guideline 4

Formulate national policy on extension in order to ensure political and financial commitment

Only a few nations—Nepal, Bangladesh, and the Philippines, for example—have developed national extension policies to date. A policy's existence is crucial because it guarantees political commitment, which determines how money is allocated. For many years, the extension profession has suffered from appallingly low compensation and operating budgets, as well as a dearth of appealing career development opportunities. This has had a negative impact on the performance of otherwise highly skilled and dedicated experts. Where decentralization or devolution has taken place, the outlook for extension is very dire. It is true that for extension to continue operating normally, appropriate policy is required to shield it from the early negative impacts of decentralization. It is imperative that extension policies be developed within the larger framework of national agricultural and rural development policies in order to bolster the institution's position and function for national development. In order to inspire and uplift extension workers, extension policy should be developed collaboratively, incorporating all relevant parties and taking into account both technical and professional development concerns. The operational ties and collaborations between extension and other pertinent service providers, such as those in the fields of information technology, agriculture education and training, farmer associations, commodities, research, marketing, and the environment, should also be included.

Key actions

1. A conference with important stakeholders was arranged, and copies of the national strategy for rural and agricultural development were distributed. Members of public and private institutions, as well as representatives from small and commercial farmers, should be included in this gender-sensitive gathering that takes place in both the provinces and the center. The following items ought to be on the agenda:

- Talk about developing an extension policy through public participation in the framework of the country's rural and agricultural development policy.
- Creation of a small committee with the deadline to design the extension policy.

2. Circulation of the draft policy document, when ready, among the participants of the first meeting for review and comments.
3. Organization of another meeting of the stakeholders for discussion of the draft extension policy and feedback
4. Finalization of the policy document by the policy draft committee members in light of the feedback and comments received.
5. Approval of the extension policy through necessary government procedure.
6. Printing of the approved extension policy document.
7. Distribution of the extension policy document among the relevant public and non-public institutions related to extension work

including the academic institutions where extension is taught as a subject, and representatives of farmers and other stakeholders.

Guideline 5

Give extension profession a long overdue status similar to other agricultural disciplines

For almost a century, Western European nations, Australia, Japan, and the United States of America have all benefited greatly from significant agricultural and rural development brought about by agricultural extension. Regretfully, the extension profession has never been given the respect it merits, that is, in comparison to other agricultural professions, despite the fact that the majority of extension organizations in developing nations were founded more than 50 years ago. Agricultural extension workers in most developing countries still face low salaries, inadequate benefits, and few opportunities to advance their careers despite the fact that agricultural extension is an intellectually and physically demanding profession. Any significant research and extension project worth looking at, especially one funded by a large donor like the World Bank, will reveal that the majority of opportunities for abroad training are taken advantage of by researchers or other agricultural professionals, leaving very few, if any, for the extension workers. It should come as no surprise that a small percentage of young men and women are eager to continue in their current line of work after high school. Until this field's working conditions are aligned with those of other agricultural specialties and the current extension organizations are provided with sufficient funding, opportunities for professional growth, recognition, and mobility—all of which are necessary for conducting proper fieldwork—suitable candidates will never be drawn to this line of work.

Despite its proven value in promoting rural and agricultural development in highly developed nations, agricultural extension in developing nations will continue to be viewed as a profession of second choice in the absence of this long-overdue change. The growing global movement to combat hunger and poverty is reinforcing the need of extension in promoting sustainable development and livelihoods. Rural areas host the majority of the world's hungry and impoverished. Given their relative frequency of visits to rural areas, extension workers have the potential to be an effective force in the fight against hunger, poverty, epidemics, and unfavorable conditions brought on by both natural and man-made disasters, given the right service conditions. To yet, though, not much has been done to enhance its working conditions.

Key actions

1. A review of the agricultural extension service's conditions in relation to other agricultural services, along with policy measures to address any major deficiencies that may be negatively impacting staff morale.
2. Meaningful participation of important extension personnel from the outset of any significant government initiative for rural and agricultural development, including those pertaining to food security, sustainable livelihoods, desertification mitigation, reducing rural poverty, etc.
3. Allocation of specialized foreign and in-country training opportunities, scholarships for degree-seeking academic programs and study-tours for extension discipline, and rigorous monitoring to verify that they are truly accessed by the extension professionals and not taken away by other disciplines.
4. Regular public acknowledgment of extension professionals' accomplishments.
5. There is a foolproof system in place to make sure that politicians or non-extension government officials don't take away the few cars intended for extension field operations for their own use.
6. Appointment of top extension personnel with experience to positions of policy-making authority for rural and agricultural development, including head of any pertinent institution.

Guideline 6

Bring pre-service education in agricultural extension in line with the modernizing of the national extension system

It is common knowledge that the majority of developing nations' higher education systems require significant development. This is especially true for agricultural extension academic programs. Undergraduate agricultural extension pre-service education is little more than lip service in terms of both duration and content. The curricula are out of date, there is little use of audiovisual aids, there is a lack of knowledge about appropriate instructional approaches, and most importantly, extension students are taught a lot of theory but receive little exposure to actual extension work that involves farms, rural areas, and farmers. At the graduate level, things are scarcely satisfactory. First off, not many students choose agricultural extension as their major due to the challenging nature of the work, the sometimes poor quality of the service, and the limited prospects for professional growth within the extension service. The idea that extension plays an equally vital part in nation-building initiatives as other agricultural disciplines and is in no way inferior to them all faces a psychological hurdle. Poor academic programs in extension produce new graduates who, when confronted with farmers—some of them for the first time in their lives—lack professional confidence and technical competence. Embarrassingly for the young recruits as well, farmers actually know a lot more about the topic these new hires are expected to provide advice on. It is true that not many extension agents are suited to live in rural areas; this is due to poor student selection for agricultural extension majors as well as a near complete disdain for hands-on training. A global movement to reform the long-standing and largely outmoded agricultural and rural extension systems is currently underway, spurred by the realization that, as a result of several global developments, extension will undoubtedly be crucial in addressing the growing educational needs of farming communities. Reorientation or in-service training for the current extension workers alone, however, won't be adequate. In the absence of reforms to pre-service education in extension, the reform and modernization of national extension systems would remain a pipe dream. Since agricultural academic schools now generate graduates who are ill-prepared to serve in a modern extension service, any genuine attempt to improve the nation's agricultural extension systems should naturally begin with revamping extension education at these institutions.

Key actions

1. The newest research on extension reforms and modernization, as well as national policy on rural and agricultural development, will be distributed, along with the terms of reference for the committee that will review pre-service education in agricultural and rural extension. The committee will be composed of representatives from national and provincial planning bodies, academic institutions, the most pertinent key government departments, non-public institutions related to rural and agricultural development, experienced, and senior extension professionals, small, and commercial farmers, etc.
2. In order to evaluate the current quality and scope of the programs against the national policy on rural and agricultural development, national development goals, and a vision of how extension services can be made most useful in practical terms to face the challenge of meeting new learning needs of men and women farmers, in line with global developments, the review committee critically examined both undergraduate and graduate education in agricultural extension at relevant academic institutions. At the very least, the following areas must to be discussed and suggestions for enhancement made:

- Curricula.
- Teaching methodologies.
- Text books and other learning materials.
- Types and extent of audio-visual aids to facilitate learning.
- Training of faculty of extension in innovative pedagogy.
- Amount of practical vis-à-vis theory emphasis, especially the period to be spent in rural areas on farms with the farmers.
- Internships under real-life conditions in the field.
- Financial assistance to extension majors.
- Linkages of academic institution with the agricultural extension department and other institutions engaged in real-life extension work.
- In-country and overseas opportunities in advanced education in extension.
- Linkages with agricultural research institutions.
- Linkages with farmers' associations.
- Realization, recognition, and emphasis on gender importance in farming and extension work.
- Gender balance in the selection of students for extension major.
- Quality and number of extension faculty and their working conditions for satisfactory teaching
- Follow-up of the graduates after a few years in service.

3. Creation of an action plan that will implement the committee's recommendations over a predetermined time frame, requesting, if necessary, financial or technical support from bilateral and multilateral organizations.

Guideline 7

Promote pluralism in extension by involving public, private, and civil society institutions

When agricultural extension was first introduced in developing nations, there weren't many significant non-public institutions or private businesses eager to participate in providing extension services. For a considerable amount of time, government agricultural extension departments had exclusive control over extension because of this. Today, the landscape has shifted, and a number of public and private entities, together with civil society organizations, are actively involved in extension or activities closely related to it. These players are willing to participate in any official initiative or method that allows them to contribute to development, sometimes for free and other times at a cost. Salespeople from various commercial firms that deal in farm inputs now visit farmers to offer assistance in addition to the government agricultural extension field worker. These salespeople not only sell their products but also provide advice on how to use them. Then, a number of field workers from NGOs and charitable organizations may be engaged in extension-related activities in rural areas. In nations where extension services have been privatized, farmers are contacted by private advising firm representatives or by individual consultants. In addition to doing research outreach activities like on-farm experiments, some researchers converse with farmers and give them a lot of extension advice. The same holds true for academic staff and students at universities who interact with farmers in order to gather data. Thus, however infrequently and randomly, many nations are benefiting from a pluralistic pattern of extension. The majority of the time, active stakeholder cooperation in the planning and execution of extension initiatives is not made possible by a well-organized framework. The government will experience a significant reduction in staffing and budgetary strain with the implementation of an organized, diverse extension planning and delivery system. Apart from the media, there are other human information sources that could be advantageous for the farmers. To protect the interests of farmers, however, a clearly defined role of government is required. This role might include policy guidance, coordination, quality control, technical support to weak commercial organizations and NGOs, and dispute resolution.

Key actions

1. Making a list of all the public and private organizations and institutions (apart from the government agricultural extension department) that provide or are interested in providing agricultural extension advice to farmers in each province and/or district, including private businesses, non-governmental organizations, farmers' associations, rural community organizations, agricultural academic institutions, agricultural research institutes and stations, etc. The following should be each institution's primary areas of focus:

- The year of establishment.
- Location.
- Technical specializations.
- Men and women professional and general services staff, with

academic qualifications and field experience, whether on temporary contracts or permanent, etc.

- Motive behind doing extension work.· Physical facilities such as transportation, office equipment, communication facilities, audio-visual aids.
- Operational linkages with relevant institutions.
- Financial aspects such as total annual staff and program budget, expenditure in previous year, sources of funding and their sustainability (to be kept confidential).
- Any major current programs or those completed in recent years.
- Does the organization work on volunteer basis or some fee is charged? If latter, details of fee.
- Is the organization willing to provide extension services on contract basis?
- Does the organization need any specialized training before it accepts contract for substantive extension work?
- Any reports and brochures about the organization.
- Geographical area of program coverage.
- Type and number of people served.
- Type of extension work the organization is interested in such as for small farmers, commercial farmers, women only, rural youth, etc.
- Any other important information that could not be covered through above questions.

2. Analyze the needs of farmers in different districts with regard to extension and training, and determine whether the current government agricultural extension service can meet those demands. If not, the option of “outsourcing” may be taken into consideration, in which the government awards contracts to appropriate public and private institutions—aside from the government extension department—for the provision of specific extension services. In this scenario, funding, and delivery of extensions are regarded as distinct functions.
3. The district extension team is trained in forming public-private partnerships, outsourcing extension work to other institutions, and overseeing the quality of the work completed. Documents for outsourcing and public-private partnerships are also prepared.

4. regular update and evaluation of the list of public and private extension service providers.
5. measures taken by the government to ensure that it effectively plays its part in the pluralistic extension pattern in order to protect the interests of farmers. These measures include policy guidance, quality control and coordination, technical support to weak private organizations and non-governmental organizations, and dispute resolution.

Guideline 8

Privatize extension partially or fully only where it is socially and economically feasible

Many important services that are still provided by the government in developing nations are now provided by the well-established private sector in industrialized nations. There is no denying that the public sector dominates in less developed nations. Nonetheless, there is currently a strong tendency towards privatization in a number of developing nations. A number of European and Latin American nations have completely privatized agricultural extension services. others economists and funders have advocated for the privatization of extension services in developing nations, and others, like Uganda, are now in the process of doing so. However, a number of recent developments, including the government reclaiming some farm support components from privatized advisory services in some developed countries, like the United Kingdom, the dissolution of public extension services in some Latin American countries at the same time that farmers were dissatisfied with alternative extension services, and uncertainty regarding the ultimate success of privatization in countries where the majority of farmers are subsistence farmers, are in fact important warning signals against rushing headlong into privatizing extension in developing countries. However, this should not be seen as a sign that privatization has failed; rather, it should highlight the need to thoroughly examine whether privatizing extension is feasible while bearing in mind the distinctions between the agricultural and farming practices of rich and developing nations. The difficulty lies in commercializing extension services where it makes sense, rather than trying to privatize all farmer services. Since most extension advice is of insufficient quality to warrant payment in cash, hundreds of millions of subsistence farmers are unable or unwilling to pay for it under the current circumstances. Commercial farmers, cooperatives, and producer associations will be more likely to pay for extension assistance if it increases farmers' revenue and is therefore worthwhile. If a partial privatization of extension is chosen, appropriate policies, guidelines, and practices will be required to safeguard the rights and interests of commercial extension advisors as well as the farmers they represent. In order to prevent farmers from being taken advantage of, the government will also need to implement quality control procedures for extension advice. Developing nations would be wise to experiment with different approaches that have been successful in other parts of the world in order to reduce the amount of money allocated to the public coffers and to stimulate private sector interest in providing extension services.

A suitable government mechanism for supervision and quality control of extension services to be provided by the private sector is one of the prerequisites for the proper privatization of extension services. Other prerequisites include the existence of competent private extension companies and individual extension advisors, strong farmers' associations with legal status that could not only effectively demand extension services but also evaluate their quality and be willing and able to pay for the advice.

Key Actions

1. A survey to determine the kind and quantity of commercial farmers, sizable cooperatives, and agribusinesses that are able and willing to use private extension advisory services on a fee basis, or hire private advisors; the greater the number of farmers who are in favour of privatizing extension, and vice versa.
2. An investigation into the existence of a sizable and potentially competent pool of retired extension workers may be conducted in order to determine whether or not there is a need for such services.
3. A research to evaluate the ability and willingness of farmers' groups, if any, to pay for extension guidance; the more associations that are willing to pay, the more compelling the case is for privatization.
4. The preparation of necessary rules and regulations and documents thereof, such as contract preparation, remuneration rates, payment methods, negotiation, and conflict resolution, a booklet containing a list of all private extension services providers in the area, with their qualifications and experience, technical areas of specialization and how to contact them, government's role in supervision, quality control and conflict management, etc, are all presumptions made by the study's and surveys that the country does in fact have a strong case for fully or partially privatizing agricultural extension services.
5. Building the capacity of individual farmers and their associations in areas like requesting extension services, maintaining records, evaluating the caliber of advice provided by extensions, etc.

6. Maintaining a suitable public extension force under government control to give free extension advise to farmers living in marginal areas and those who are unable or unwilling to pay for such advice.

Guideline 9

Develop and apply information technology tools to facilitate the work of extension workers

Extension organizations may profit from information technology advancements in the same way that other types of organizations gain from the development of various technologies. When electricity is available, the extension offices can use e-mail, the Internet, and a variety of cutting-edge audio-visual tools in their everyday work. However, in practice, the benefit will be restricted due to the fact that the majority of rural areas in developing nations lack traditional telephone infrastructure and energy, with the exception of cell phones. There are many who advocate replacing extension agents with information technology centers located in the villages.

This hardware- and technology-driven strategy is unhealthy because information and communication technology should be used to create tools that extension workers could use. It is common knowledge that in certain nations, farmers have harmed their crops due to misinterpretations of technical advice given over the radio. This was especially true with regard to the suggested dosages of chemical insecticides and pesticides. Subsequent studies revealed that farmers who participated in radio listeners' groups, or other programs supported directly by extension personnel, genuinely profited from the extension messages broadcast on the radio.

Key actions

1. Because local conditions vary, it is challenging to recommend a certain information technology application for a given nation. Nonetheless, the following is offered in a general sense that could be modified for specific circumstances. Generally speaking, the best places to begin implementing information technology are those with the required infrastructure and prerequisites. Naturally, the other sections will have to wait. A team of two experts, one in electronic communication technology and the other in extension and training, will carry out the following research:

- Identify specific activities in extension work where modern information and communication technology may be realistically applied.

- Identify the extension work related locations in the country which have necessary infrastructure for installation and use of electronic technology tools.
- Prepare a plan for developing and installing necessary tools.
- Prepare a plan for training of the staff that will use those tools.
- Prepare cost estimates for the entire work.

2. Extension and pertinent institutions may develop interactive electronic links. One such is the Virtual Extension Research Communication Network, or VERCON, a tool created by the Food and Agriculture Organization (FAO) to improve the operational links between research and extension. These linkages are still inadequate in the majority of developing nations because of, among other things, physical distances and subpar conventional communication methods.
3. Extension databases could be established with data on the prices of commodities over the past few years and forecasts for the near future, climate records from the previous few years and any unusual weather conditions anticipated in the near future, agricultural technologies that have been proven to be beneficial, contact information for subject-matter experts and buyers of agricultural produce, demand for grains and vegetables, etc. The extension workers may utilize this database in the course of their job.
4. A computer and inventive methods can be used to create an array of visually appealing extension and training materials.
5. To ensure that there is a human follow-up available on a periodic basis, tele centers may be developed in the areas that are typically unserviced by extension workers due to factors like remote locations and staffing shortages.

Guideline 10

Develop original, location-specific, participatory, gender-sensitive, and inexpensive extension methodologies and materials instead of applying those methodologies which are promoted as universally suitable

For many years, the majority of developing nations have adhered to top-down extension approaches. Another top-down extension model that was later criticized for a number of reasons was the Training and Visit (T & V) system, which was established by the World Bank in the 1970s and dominated extension work in many countries for more than 20 years. Farmer Field School (FFS) is another significant and well-liked methodology. It was first applied to FAO-supported integrated pest management (IPM) initiatives in Indonesia and the Philippines, where it became the industry standard. Because the FFS placed more of an emphasis on student participation than the T & V system, it was chosen. In essence, both of these extension approaches were not truly created locally, and despite having some positive aspects, they are said to have the flaw of being excessively expensive, particularly when it comes to upscaling, which calls into doubt their long-term viability. Unfortunately, there hasn't been much enthusiasm for developing any situation-specific methodologies, aside from a few modest but appreciative attempts to experiment with new extension methods under some projects, like those funded by certain bi-lateral West European donors. This is because developing countries don't appear to have gone beyond the two main extension methods mentioned above. The T & V system is currently out of fashion, and the majority of nations have either abandoned it completely or significantly altered it such that just a few details remain from the original design. The FFS approach has been criticized for a number of reasons, yet it is still being adopted in some nations because to the support of specific donors. Although the FFS is a good methodology for teaching farmers technical subjects related to integrated pest management (IPM), there is concern that, like the T & V system, it may ultimately suffer from the same fate as the T & V model due to the strong push by its sponsors for its universal application for all technical subjects and geographical regions. Certainly, there is a compelling case to be made for creating location-specific extension techniques. It is not logical or technically sound to try to apply any one extension methodology everywhere. Despite funders' insistence on its adoption, no single extension methodology—no matter how successful it has shown to be in a particular situation—can be acceptable for all scenarios. This fact has been confirmed via field observations and experience over many years in various nations. When creating an appropriate extension methodology, it is necessary to take into account the following factors: people and their characteristics; farming patterns; geographical terrain; climate; population density; type and diversity of extension service providers; political environment; institutions; infrastructure; local customs; and possibly some other factors. As a result, a nation like Nepal, which possesses plains, high mountains, irrigated areas, and rain-fed areas, would naturally develop a variety of extension techniques to address each distinct set of conditions. The T & V extension system, implemented across Nepal as part of World Bank-financed projects, actually performed very well in the irrigated plains where there were advantages such as improved technologies for growing wheat and rice and good transportation for extension workers. However, the technology available in the western rain-fed region was woefully inadequate, and the high mountainous areas nearly failed because of the extremely limited mobility of farmers and extension workers due to unfa In a similar vein, those who farm in desert regions or on islands will employ different extension techniques. When developing an effective extension approach, it is important to consider the type of farmers involved, including whether they are subsistence, commercial, male or female, physically fit, or incapacitated for other reasons (e.g., HIV/AIDS).

Key Actions

1. Assembling a small group of seasoned extension workers with experience in the field to take on the following duties in order to create novel, context-specific extension methodologies:

- Through extensive and ongoing consultations with men and women farmers, a variety of traditional and modern free or inexpensive non-formal educational methods for a range of technical and non-technical subjects and skills have been identified. These methods, which are primarily intended for illiterate men and women and do not adhere to the formal education system's criteria, have evolved over centuries and are widely used in rural areas across the nation. An emphasis has been placed on these methods, which prioritize exposing students to real-world experiences.
- Application of the principles of client participation, gender-sensitivity, bottom-up program planning and decision-making, and client-focus to the major non-formal education methods identified while keeping in mind the situational factors of each location (people and their characteristics, farming patterns, geographical terrain, climate, population density, type, and diversity of extension service providers, political environment, institutions, infrastructure, local culture and customs).
- The reformulation of diverse non-formal education approaches based on principles into organized extension methods and the creation of pertinent audio-visual resources that could supplement instruction.
- Over a few months, men, and women farmers from different agricultural interest categories will test each extension methodology in the field using pertinent training audio-visual materials. At the end of the testing process, the extent to which the farmers have learnt to apply their new knowledge and skills in the real world will be evaluated.

- Naming Each extension technique after a unique, appropriate local term.
- Identifying basic, culturally-based patterns for dividing different farmer categories into gender-sensitive groupings at various locales for educational objectives by using the most appropriate location-specific extension techniques.
- Creation of a handbook on diverse, unique, site-specific extension techniques, complete with audiovisual resources, for a range of farmer categories, to be utilized by extension agents in disparate locales.
- Creation of a training program and implementation schedule to teach extension employees how to apply extension approaches in practical settings and modify them to fit the unique circumstances that arise.

Guideline 11

Orient extension staff to major food security related global developments that could eventually affect rural livelihoods

The majority of developing nations do not currently see a strong need to educate their farmers on issues pertaining to globalization, market liberalization, genetic engineering, and biotechnology, all of which will eventually have an impact on their communities. This is because farmers generally believe that these developments at the global level don't directly affect them and are too complex for them to understand. Although for the time being this opinion may be correct, it won't be long before these advancements begin to have an impact on rural residents' means of subsistence. For instance, the recent admission of China to the World Trade Organization (WTO) would undoubtedly force a thorough review of the nation's current laws and practices on agricultural production, agro-processing, storage, export, marketing, and quality control. Because of this, China's national extension system needs to be prepared to begin teaching farmers about the potential consequences and expectations of the nation's WTO membership. Amidst market liberalization, farmers will also have to deal with the problem of farm subsidies. Actually, it has already begun. For instance, small farmers in Mali who have been cultivating cotton for many years are facing significant challenges as a result of their inability to match the extremely low price of imported cotton, which the foreign farmers were able to produce at extremely low cost because of significant farm subsidies that they received from their respective governments. Until now, academics, and extension agents in developing nations have paid little attention to the environmental aspects of agricultural innovations. Almost all national agricultural extension services still have a very specific mandate, which is restricted to the transfer of agricultural technology. Extension agents will vigorously support a new technology without considering its environmental friendliness as long as it can increase yields. Aside from strong extension assistance, the majority of the Green Revolution's achievements throughout the 1960s were attributed to high yielding variety development and cultivation, appropriate irrigation practices, and the heavy use of pesticides and fertilizers. That recipe for getting large yields is, however, currently being criticized for its disregard for environmental preservation and the preservation of natural resources. For this reason, there has been positive attention given to technologies such as IPM. Serious attention must also be paid to population increase, since it counteracts any gains in agricultural productivity. For instance, Asia's population grew by 60%, or more than one billion people, between 1970 and 1995 despite a drop in human fertility in most other regions of the world. Despite the strong correlation that exists between agricultural productivity, population, and food security, very few extension agencies have focused on enlightening farmers about this crucial topic. While it might seem early to pose these kinds of enquiries, farmers will soon be asking extension workers about genetic engineering and biotechnology. Extension agencies must to begin creating basic instructional materials and initiatives to increase competence in these fields.

Key Actions

1. Provide brief orientation and training programs in the short term to train extension staff at all levels on topics such as market liberalization, farm-market-chain links, quality production and market of exportable commodities and the kind of robust support that should be given to producers; other topics include environment sustainability, natural resource management, genetically modified crops and vegetables, and organic farming.
2. After giving the relevant training to the stakeholders involved, create, and implement extension approaches such as FAO's Agriculture Population Environment Extension (APEX), which incorporates population and environment educational messages into ongoing agricultural extension initiatives.
3. Update pre-service education in a number of areas as a long-term solution, following the guidelines outlined in the guideline "Bring pre-service education in agricultural extension in line with the modernizing of the national extension system."

Guideline 12

Encourage the extension services to empower farmers through organizing them into legal associations to constitute a strong lobby for themselves and for extension

There is a global trend towards turning farming into an agri-business or commercial venture. There are economic arguments against the continued practice of subsistence farming. In many developing nations, farming has been passed down from generation to generation as the younger generation does not perceive much opportunity in this field. Other worldwide trends, like those already discussed in this paper, also suggest that farmers need to diversify their sources of income in order to be less dependent on farm income.

Without a certain, the farmers will never be able to manage their tiny farms as an agri-business or commercial venture, nor will they be able to build a powerful self-interested lobby.

This is especially true if they operate individually rather than collectively. They must first combine themselves into informal, tiny organizations, then progressively develop larger groups with specific interests, and then register as legal companies with the government. Strong farmer advocacy will not only fight for farmers' rights but also push the government to provide efficient agricultural extension services, particularly in nations that have experienced decentralization or devolution. This will determine the amount of funding and personnel required for the extension services. Moreover, farmer groups will make it easier to use participatory extension strategies. All in all, these advantages will hasten the development of agriculture and rural areas.

Key Actions

1. Give extension agents refresher training so they can help men and women farmers organize special interest groups.
2. Provide farmers with direct extension services so they can form associations and groups and register with the government to obtain legal status.
3. Give farmers' associations the training they need to gain the knowledge and abilities needed to use their clout to demand certain services (like credit, farm inputs, marketing issues, and extension advice from public and private institutions, and evaluating the quality of such advice when delivered), as well as to turn their farms into agribusiness and commercial ventures.

Guideline 12

Encourage bottom-up, grassroots extension program planning by farmers in order to make extension demand-driven, but also exercise supply-driven, top-down modality for promoting common public good practices such as conservation of natural resources and environment protection

It is important to support both male and female farmers in organizing groups and creating plans for demand-based extension services with the help of extension agents. These services can be fully supplied by the private sector, fully by state extension services, or by a combination of public and private entities. Nonetheless, since farmers can view these services as low priority, using an entirely demand-driven approach to services will not be very sensible if the goal is to promote public good practices. This is a result of the overall lack of awareness or disregard among farmers for widespread public welfare initiatives like environmental conservation and resource preservation. In order to promote sustainable development, the government's approach to these common public goods should be top-down and supply-driven. A common pasture, utilized by numerous rural residents to graze their cattle, is an example of the common public benefit. However, nobody seems to care about maintaining and replenishing the pasture's fertility. The herders just relocate to other lush grazing areas when overgrazing depletes the pasture. Legislation governing the appropriate use and replenishment of these shared grazing areas should be considered by the government, and any extension effort in the region should include upkeep as a fundamental component.

Key actions

1 Extension agents are trained to act as facilitators for farmers' groups, assisting them in creating strategies for demand-for-services extension programs and referring them to subject-matter experts for assistance with complex technical challenges.

2 Training of rural community leaders in grassroots bottom-up planning for creating extension service demand-for-services plans, assisted by extension workers.

3 Combining the demand-for-services plans of many groups into village plans and multiple villages' plans into cluster demand-for-services plans.

4 Integration of government district growth plans with cluster demand-for-services plans, ensuring that the district extension program

includes the implementation strategies for government directives on the use of shared public goods.

5 provision of the services through outsourcing agreements based on demand-for-services plans, either by government line departments or by non-public organizations.

6 Together with the help of subject-matter experts, farmers' groups and extension agents will monitor the promptness and quality of service delivery and evaluate the impact.

Guideline 13

If the extension function is to be performed with relatively small number of extension staff, follow appropriate strategies for getting maximum output

The structural adjustment initiatives led to a significant reduction in the number of human resources in public extension departments in a number of developing nations. Some public extension programs have come under fire for being incredibly resourceful but lacking in efficacy. The optimal number of extension workers cannot be determined by a single formula; instead, it must be determined locally, taking into account a variety of factors including geographic terrain, farming population density, infrastructure, transportation options, accessibility to electronic media, and maybe more. Given that the information technology revolution has given extension organizations additional avenues for connecting with farmers and other institutions, the outdated notion that each extension agent should cover a specific number of farm families should now be modified.

Key actions

1. The following strategies, alone or in combination, will assist in addressing the staffing shortfall in providing extension services. In this publication, some of these have already been explained elsewhere.
2. Try using the "outsourcing" option if money is not a major limitation.
3. Utilize media and information technology tools, but ensure that there is sufficient human support.
4. Examine and update the job descriptions for different staff categories, and combine positions where job descriptions overlap or are very similar.
5. As part of the "pluralism" mode, assign roles for delivering extension services to both public and private actors in civil society, including

non-governmental organizations. Ensure that the implementation is adequately overseen to ensure quality control.

6. Instead of employing solely for permanent employment, think about hiring temporary employees on contracts or on an as-employed basis.
7. Determine which farmers are progressive and who, in addition to extension professionals, could serve as facilitators for extension operations.
8. Reduce the number of employees at the national level if devolution is not being practiced, and at the provincial level if it is, by moving the majority of extension personnel from the provincial level to the sub-district and village levels. This will leave the top staff strength at a small secretariat.
9. Handle extension delivery and financing as two distinct tasks.
10. To the greatest extent feasible, avoid individual farmer contact by using the group extension technique.

Guideline 14

Ensure effective operational linkages between extension and research and other key relevant institutions

Extension is a service designed with farmers' best interests in mind. Its original purpose was to impart improved agricultural technology, but it has now expanded to support sustainable rural and agricultural development. Since its official introduction in developing nations, the most significant operational linkage that has been highlighted for extension is with research. Given that agricultural research institutes serve as extension organizations' primary source of technology, this connection is crucial. However, it is a painful fact that this specific correlation has consistently remained weak, making it a matter of great academic and practical interest to relevant specialists for a considerable amount of time. Over the past 50 years or so, a number of strategies have been attempted to improve collaboration between research and extension. A tactic employed by certain nations involved placing the two disciplines organizationally under the same leadership. The goals were admirable, but regrettably, research severely overshadowed and undermined extension, regardless of whatever developing nation chose this path. Almost always, the majority of the funds allotted were used for research. Even in large donor-financed initiatives, it was frequently noted that, despite the project design's emphasis on a fair division between extension and research, a disproportionately high percentage of foreign fellowships were stolen away by research. Establishing tiny extension units within research institutes was another tactic used in certain nations to strengthen the relationship between extension and research. However, these units only published pamphlets and booklets about the research conducted by those specific institutes, which was primarily done for PR purposes and to make the researchers feel proud of their work. These units undertook outreach efforts for research in a few unique circumstances. With virtually no prospects for advancement among researchers, the lone extension (and occasionally media) specialist in charge of the unit found himself cut off from his peers in the field. This tactic was unsuccessful. The practice of requiring researchers and extension workers to meet before the main crops seasons began was another tactic implemented under World Bank-financed initiatives that utilized the Training and Visit model of extension. The meetings' objectives included talking about anticipated field concerns, identifying the most important messages for field extension agents to convey to "contact farmers," and other matters pertaining to the upcoming cropping season. It was discovered that this modality was helpful in giving the experts from research and extension a fruitful forum to talk about the issues and worries of farmers. Similar to the Training and Visit approach, these meetings were beneficial, but they too proved to be unsustainable and ended as certain initiatives came to an end. Under the Applied Research Management II Project, which is funded by the World Bank, Indonesia has established a very promising framework for coordinating research and extension in the previous few years. Several provinces now have Agricultural Technology Assessment Institutes because to the project. After numerous minor research stations and extension information centers were shut down, they hired researchers and extension professionals to work for them. According to a ministerial directive, every employee has the same job description and benefits. Senior extension specialists or senior researchers oversee the institute. On 50-hectare plots, the institutes work with farmers to field test the technologies that are given to them by central commodity research institutes. If the technologies show promise, the district extension officers work with farmers to test them again on 500-hectare plots at a cost-sharing arrangement. If the technologies continue to function well, the extension service will begin to scale them up. By means of the "competitive research grant" mechanism, these institutes work not only with public research institutes but also with university research faculties and private firms. As of right now, this paradigm offers the best chance for research and extension to collaborate closely and with full participation from farmers, the extension service, and other stakeholders. As information and communication technology has advanced, initiatives to bridge the gap between research and extension have begun to capitalize on these developments, and virtual connections are being formed between pertinent institutions.

Key actions

1. The past approach of placing extension and research under the same administration should not be repeated because it has weakened extension. Rather, separate administration should be used for research and extension, and the following actions should be taken to ensure that the two disciplines are coordinated.

- To provide policy advice and encourage coordination at all levels, form a joint committee with senior officers from research and extension at the national or provincial level, depending on whether these services have been decentralized or not.
- At the district level, appoint researchers as subject-matter experts to provide technical support to extension agents. Such appointments, if research has not been decentralized, ought to be made for a duration of roughly three years, during which the district extension officer should serve as the primary supervisor of the researchers (subject-matter experts). A researcher should work for his parent research institution for roughly 30% of his time and support field extension efforts for 70% of his time. The relationship that these researchers have with their parent universities is crucial since it keeps the researchers' technical ties intact, counts their experience, and ensures that they don't miss out on opportunities for advancement within their respective institutions. These researchers will be regarded as resource persons when they return to their parent institutions because of their first-hand field experience, which has allowed them to share professional ideas with their colleagues. These researchers ought to be replaced by other researchers on a rotating basis. Since most researchers dislike having their assignments made at the district level, specific field allowances ought to be taken into account for these placements in order to foster motivation.
- Start a procedure at the field level for developing a research agenda by consulting with farmers, extension agents, and researchers. The goal is to make the agenda focused on field problems.
- Establish virtual, interactive connections between research and extension institutes by utilizing contemporary information and communication technologies.

Chapter **7**

CASE STUDIES

Contents

7.1. OVERVIEW

Some of the factors to take into account in this case study chapter are presented as questions:

1. Why was change necessary or desirable?
2. What situation or events led up to the reform, innovation or development that constitutes the core of your case study?
3. What were the innovations or reforms introduced?
4. How did the reform, innovation or development evolve?
5. Who delivers the services being provided?
6. Who pays for the services being provided?
7. Who administers the services being provided?
8. What specific services are provided? What is delivered?
9. What type of information?
10. How are the services provided?
11. What methods are used?
12. Do we use face-to-face, media, or electronics?
13. What have been the results so far?
14. In general, does the reform and innovation affect rural development and poverty alleviation?
15. What, if any, are the impacts on the socio-economic situation of the service recipients?
16. How do policymakers and stakeholders view the extension services?

The purpose of the case studies was also to demonstrate the significance of extension changes, as well as their potential for sustainability and replication. Because changes are sometimes very new, there is often little proof of their impact; as a result, the case studies' approaches to the challenges vary. Impact, sustainability, and replicability are ultimately the main concerns that shape the studies' main points of emphasis.

7.2. SPREAD OF LOCAL AGRICULTURAL RESEARCH COMMITTEES

In Colombia, a Local Agricultural Research Committee (abbreviated CIAL) (Figure 7.1) is a group of individuals who volunteer to conduct studies in rural areas on behalf of their customers. The committee's client group can be an agricultural enterprise, a rural community, or an interest group like a producer organization or women's group. By sharing the knowledge, expertise, and advantages that come from experimenting together with the associated costs and risks, CIALs support equitable rural innovation. When a group is inspired to start a CIAL by speaking with a facilitator or learning about the approach from other farmers, that is the first stage in the formation process. The group then gets together to choose a committee, lists issues and possibilities, assigns a priority to each, and instructs the committee to conduct experiments on their behalf. The committee then creates trials to fulfil this need. The CIAL approach mandates that trial plots begin modest, which lowers the possibility of financial loss in the event that studies fail. Furthermore, the approach lowers the possibility that the committee will suggest an unsuitable technology by mandating that every trial be duplicated and that successful trials be repeated across three seasons on increasingly expansive storylines.

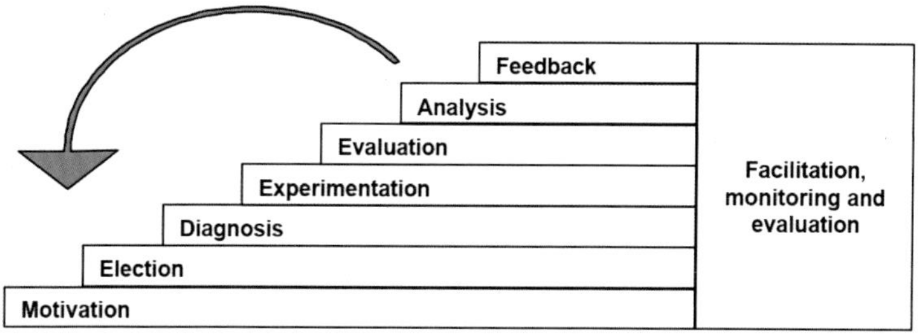

Figure 7.1. *Steps in the CIAL process.*

Source: Alex and Rivera, (2004).

The CIAL approach was created in the early 1990s by a team at the International Centre for Tropical Agriculture (CIAT, to give it its Spanish name). The group had spent more than five years trying to figure out why farmers with limited resources, in particular, were not utilizing the technology developed by official sector research. They came to the conclusion that farmers needed to be involved sooner in the development, testing, and local adaptation of new technology if adoption rates were to rise. They understood, however, that doing this in the intricate and dangerous ecosystems that millions of farmers depend on would be very expensive and time-consuming unless agricultural communities took the lead. In order to support farming communities in doing their own on-farm evaluation and adaptation, the team created the CIAL technique. As part of the CIAL approach, farmers are expected to become knowledgeable about and use the notion of experimental replication in order to help formal research and development institutions better understand the requirements of impoverished rural communities by using their findings. CIAT got its start in 1990 when it established five CIALs in Colombia's Cauca Department thanks to support from the Kellogg Foundation. The CIAT team created 18 CIALs in all by late 1991; by 1994, that number had increased to 55. More than 250 CIALS are currently in operation in eight Latin American countries, and there are an undetermined number of other countries in sub-Saharan Africa and Asia, including China, that have adopted the strategy. CIAT has also trained trainers from other nations and other organizations that went on to establish their own CIALS. In 2002, nongovernmental organizations funded 57% of CIALs that were known to exist, while government organizations supported 33% of them. The remainder was facilitated by consortiums made up of two or more collaborating organizations. Most communities prioritize their primary food crop research during the CIAL diagnostic procedure. In the 250 CIAL communities in Latin America, maize, beans, and potatoes account for 59% of crop research. As a result, while communities prioritize potatoes and broad beans in the Andean regions of Ecuador and Bolivia, the majority of CIALs in Honduras work with common beans and maize, which are the two main components of the local cuisine. A wider range of subjects are covered by CIAL research in the few places that have good food security. Committees work to increase revenue in these circumstances by pursuing new agricultural ventures or enhancing the value of established ones through enhanced processing.

7.2.1. Benefits and Costs

A CIAT impact assessment of the 68 Colombian CIALs in 1998 found that:

- CIALs directly resulted in more rapid technology adoption,
- CIAL process itself has led to people learning useful skills and forming valuable social linkages, and
- CIAL communities had experienced improvements in welfare.

People began establishing agro-enterprises in response to the experiment outcomes, as well as the new connections and abilities they had acquired, which contributed to advancements in welfare. As an instance, a few CIALs have begun to commercially manufacture the seeds of the finest crops found during their studies. Compared to farmers in comparable villages without CIALs, CIAL communities have more crops and more kinds in their fields as a result of the experimentation that has motivated them to try more novel crops. Villages with CIALs are better able to manage risk because of this diversity. Furthermore, the poorest farmer strata were equally likely to adopt as the wealthier strata, and technology was adopted more quickly in communities having CIALs. As a result, CIALs facilitate the quicker adoption of better varieties by communities, regardless of whether they are created by farmers or the official research sector. Because farmers in villages with CIALs reported fewer "hungry months" of seasonal food shortage, the study further implies that CIALs may increase food security. Women's participation in local research has been one of CIALs' major effects. Since women make up approximately 60% of committee members as of 2003, farmer evaluations now take into account important factors like cooking time and taste, which can determine whether or not a community embraces new technology. It is now possible for women to form their own CIALs (fourth of CIALs are exclusively for women) to do research on issues that are important to women, like family nutrition. Women have benefited monetarily from CIAL research, increasing family earnings in the process. CIALs have also had an effect on formal sector research objectives. For instance, since 1996, Ecuador's national research and extension organization, INIAP, has supported 19 CIALs in one of the country's five regions. INIAP workers are now placing less attention on potatoes and more on crops that farmers are interested in, such as the native quinoa (Chenopodium quinoa), beans, and chocho (Lupinus metabolism), after learning that resource-poor farmers prefer to diversify their crops. The positive relationships that INIAP staff members have built with communities as a result of the CIAL process also serve as a source of motivation for them when working with CIALs. While CIALs have an impact on INIAP's research agenda in Ecuador, this is more of an anomaly than the norm. Overall, the connections between CIALs and formal sector research are not as strong as first anticipated; further research is required to determine the reasons behind this and to explore potential ways to improve the connections. CIALs do have costs, even though they were intended to be an inexpensive means for a research and extension agency to increase its reach. According to Ashby et al. (2000), the primary expenses were educating the facilitators who aid in the process and giving the CIALs a little research fund. From 1990 to 1998, the expected cost of establishing a CIAL was US$670 for the first year and US$325 for the following five years. A cautious estimate of 78% was made for the return on this investment. However, because of "learning by doing," the costs of establishing and maintaining CIALs have significantly decreased, meaning that the return on investment is probably much higher currently. For example, it has been discovered that one facilitator may support up to 50 CIALs under the correct circumstances, and that experienced farmers can competently teach facilitators for less money than can hired specialists. Today, the initial start-up fees for each CIAL might vary from $25 to $500, either in cash or in kind (Ashby, 2003).

7.2.2. Sustainability and Replicability

CIALs are dynamic organisms. Following the completion of the first study cycle, some CIALs will start a second cycle to look into a new issue or opportunity, while other CIALs will stop their investigation and might begin marketing a portion of the newly tested innovations. For instance, a CIAL in Cauca, Colombia, discovered a bean variety with a high yield and over the course of the next seven years, produced 230 MT of seed before the variety developed a susceptibility to the fungus anthracnose. Now, the CIAL is starting a second research cycle to find new bean kinds, including climbing beans for the first time. Whether or not CIALs are held, the CIAL process enhances the community's ability to experiment and find new ideas in the long run. The establishment of agro-enterprises, experimentation, and proactive search for answers are essential for rural communities to remain sustainable in the current global environment of climate change and increasingly open markets. When CIALs were first established, one of the concepts was for the committees to serve as a means of feedback to the National Agricultural Research and Extension Systems (NARES). Since then, NARES in Latin America has been severely undermined by funding reductions. The public sector's role in paying non-formal agricultural extension, albeit not necessarily in providing it, has just come to light, therefore the pendulum may be swinging back (Rivera, 2003). Mature CIALs can reach impoverished and remote client groups with low-cost research and extension services, according to experience with them. CIALs might be well-positioned and gain from increased funding to NARES from the public sector. Evidence from Ecuador, Colombia, and Bolivia does, in fact, indicate that CIALs are "mainstreaming" at this time. Bolivia has restructured its NARES, requiring rural communities to incorporate farmer opinions into their municipal development plans as a matter of law. CIALs are showing to be a helpful tool in achieving this. INIAP has reorganized in Ecuador to focus on organic agriculture through participatory techniques. In order to maximize impact with the limited resources at its disposal, INIAP has realized that research and extension that ignores farmers' needs and experiences might be "like throwing money in the river"1. Participatory methodologies, in particular CIALs, are essential.

Since implementing the technique in 1996, the Colombian national research program CORPOICA has established 46 CIALs across seven departments. In 2001, CORPOICA was collaborating with 30 CIALs. According to a case study, the institutionalization process of the CIAL methodology is hindered by the commonly held belief that it is merely an extension tool and not relevant to scientific research, even though the methodology has official backing within CORPOICA (Menter, 2002). However, among the scientists who have participated directly in CORPOICA, the CIAL methodology has gained traction. Creating so-called "second-order organizations" has been another method of guaranteeing CIAL longevity. In order to safeguard and advance their interests, the CIALs in Cauca, Colombia, established CORFOCIAL as an umbrella organization in 1995. Operating on the income from an endowment donated by an unidentified patron, CORFOCIAL employs three paraprofessionals. It facilitates visits to research facilities and other CIALS, offers training, assists in crafting funding submissions, encourages the sharing of goods and seeds amongst CIALs, and plans the annual CIAL meeting. In Honduras, a federation of CIALs was formed in 1998 with financing from the Participatory Research in Central America (IPCA) project 2. Like CORFOCIAL, this organization (Association CIAL--ASOCIAL) receives funding from an endowment and performs comparable tasks. Additionally, members of ASOCIAL and individual CIALs have access to savings and microcredit programs. Another distinction with Honduras is that researchers from national agricultural programs frequently attend the annual CIAL meeting, indicating a potential future improvement in the country's integration of CIAL and formal sector research (Humphries et al., 2000). The general absence of official connections to research and extension organizations implies that CIALs' financial viability is a problem. This is partially a result of the difficulty that all community-based organizations are facing as government funding for agricultural extension and research evaporates. CIALs have created a variety of methods to reload their operating fund. If CIALs are to continue, these local efforts most likely need to be matched by some outside assistance. 12 years of working with CIALs have demonstrated that three fundamental ideas are the foundation of success:

- Relationships between the CIAL, the community, and external actors are founded on mutual respect and accountability and shared decision-making.
- Partners in the research process share the risks of research.
- Research is conducted by systematically comparing alternatives.
- Knowledge is based on building experience and learning by doing.
- Research products belong to the community.
- Another key factor is adequate training of CIAL members in the participatory research process.
- CIALs tend to be more successful in communities where social capital is already high (Humphries et al., 2000).

7.2.3. Lessons Learned

The fact that CORFOCIAL and ASOCIAL are independent organizations that prioritize the needs of its members is one of their strengths. The methodology's potential has been demonstrated by the CIALs affiliated with these two organizations. But in order to "mainstream" the strategy, more CIALs must be developed in already-existing organizations. Based on past experiences, it is imperative that NARES staff adhere to the fundamental tenet that the primary purpose of a CIAL is to serve the community to which it belongs, rather than the NARES's goal in adaptive research or extension. In addition, regular communication, responsibility, respect for farmer research, and cooperative decision-making are required of NARES staff members. Mentor (2001) offers the following suggestions for institutionalizing the CIAL approach, based on the CORPOICA case study:

- Find your natural supporters first, that is, establish a foundation of support before trying to persuade doubters.
- Make a demand for training by using the success data that is currently available.
- Utilize suitable media for various audiences to increase outcome awareness.
- Assist important parties in choosing how to collaborate with farmer research committees.
- To improve institutional learning, implement participatory evaluation and report-back at all levels.
- Pay attention to the outcomes as well as what may be learnt from the CIALs' procedure.
- Once CIALs are formed, gradually reduce the amount of time researchers spend working with each group; utilize a schedule that reduces the frequency of visits.
- Interact with both seasoned professionals and those who are just starting out in CIALs to encourage process growth and share ideas for tweaking the methodology. While adherence to the fundamental CIAL principles outlined above is crucial, it's equally critical to support regional variations. Some effective modifications include: where ensuring food security in the short term is of utmost importance, start by assessing treatments in trials conducted by researchers, and then share risk in more uncertain kinds of farmer-run experimentation (Ecuador, East Africa).
- Combine the CIAL's small experimental plots investigating hazardous technologies with a group production plan utilizing proven technology. The group's output increases the petty cash fund and helps reimburse committee members for their time (e.g., Honduras, and Colombia).

- Without setting up official tests, test, and track innovations on farms; this is especially helpful when it comes to livestock or natural resource management techniques (East Africa and South-East Asia).
- Choose a sizable committee. Large committees in Northeast Brazil helped CIALs get through times of seasonal migration by replacing migratory members with those who remained. Large committees in Honduras allowed a wider range of the clientele to gain from human resources.
- Create a CIAL to offer small agro-enterprises, whether they are new or established, R&D on new goods or processes.
- Establish a petty cash fund by giving the CIAL experimental inputs in kind. The committee's operations should then be funded by trial earnings. This increases the petty cash fund of CIALs in Bolivia and Colombia.
- Manage the petty cash fund as a small venture capital fund that offers loans for equipment that is rented out to the clientele, or as a revolving credit fund.

Based on empirical evidence, CIALs can evolve in two ways: either they stay involved in community-based volunteer research or they turn their research findings into a profit-making agrobusiness. To make sure the CIAL stays in touch with the community and takes the first path, it is crucial that they have regular meetings where they report back to them. But CIALs that establish agro-enterprises also help the society by, say, supplying seed for novel and reliable crop kinds. One of the conclusions, in fact, is that the CIAL technique works incredibly well for kickstarting the development of agro-enterprises. Lastly, it has been demonstrated that CIALs are an excellent complement to Farmer Field Schools (FFS), which have the capacity to create a local agro-ecological knowledge base that enhances the significance of CIAL research.

7.3. CATALYTIC ACTION FOR THE EMERGENCE OF FARMER EMPOWERMENT

This case study looks at the International Support Group's (ISG) "catalytic action" to empower farmers in advance of the successful adoption of "demand-driven" extension policy reforms. The first benefit of the catalytic action is the empowerment of farmers to band together and share a vision for their village's future that takes into account the natural resources that surround it. The next step of the initiative is to give farmers the tools they need to manage their own multi-stakeholder learning process independently of traditional project formats and without the aid of outside funding sources. The catalytic action is a "one time" brief intervention that is carried out in three stages: capacity building in multi-stakeholder collaborative learning processes; second, self-organization among interested stakeholders; and third, exploration of interest among key stakeholders. The catalytic action is carried out with the explicit assumption that no money will be left over for any kind of action by farmers, their local leaders, or service providers. All initiatives are "self-help" in the sense that they rely on their own resources and ongoing employment. Danida's request to ISG to investigate local interest in using multistakeholder collaborative learning processes to support farmer demand for agricultural support services resulted in this catalytic action. Danida wasn't the only "stakeholder" who wasn't happy with how the revisions to the farm extension policy were going. Farmers were unhappy that they were not receiving the complimentary inputs and training that they were accustomed to. Extension agents were unhappy that they were not receiving enough funding to visit farmers and carry out their regular programs. Local governments were dissatisfied since "their" extension workers weren't putting in a lot of effort for them because NGO's or party politics were using them. It proved more difficult to begin "demand-driven" extension than planners and lawmakers had anticipated. An order from above cannot empower farmers. Instead, self-awareness, self-organization, and group action lead to empowerment.

7.3.1. Catalytic Action for Farmer Empowerment East Africa

Through "demand-driven" extension, farmer empowerment emerged as a result of the catalytic activity.

- Helping farmers to articulate "common" future visions of their community and demand services,
- Helping farmers negotiate with public and private service providers for the necessary services,
- Changing the relationships between key stakeholders such that they can all collaborate to help farmers realize their future visions, and
- Facing up to the implications this has for the way each organization worked.

Following the capacity building workshops, Danida's financial assistance for the "catalytic action" ceased. Every group has been using their own resources ever since. In a similar vein, ISG has maintained contact with the multi-stakeholder groups that arose via impromptu meetings, email correspondence, and participation as training resource persons. All of the effects listed here were accomplished by locals banding together on their own initiative and using their own resources. There is no reason to think that they won't carry on doing so indefinitely. During the capacity building workshops, there are participants from the village, district, and national levels who engage in catalytic action. Participants in the workshop included:

- Leaders of farmer groups, elected officials, local government representatives, educators, and extension agents at the village level.
- Public sector representatives, elected officials, and local government officers in the fields of agriculture, community development, and planning, as well as representatives from the private sector, including NGOs, cooperatives, farmer unions, input suppliers, credit, and marketing agencies, are present at the district level.
- Academics, researchers, development professionals from government ministries of agriculture, forestry, environment, and local government planning, as well as non-governmental organizations, cooperatives, and farmer unions, are involved at the national level.

The diversity of participants in the program facilitated two outcomes. Initially, it assisted farmers in involving the village leadership in the unofficial "learning groups" they established upon their return home. Secondly, it offered a district-level informal "learning group" that the peasants could connect with. For sustainability and scaling up, these connections between organizations and village leadership, as well as between village and district, are crucial.

7.3.2. Impact

Following their return home, participants engaged the village leadership and formed "multi-stakeholder groups" to mobilize the villagers to negotiate partnerships with a wide range of service providers and local government officials. The results of their efforts are described below.

1. Kenya

i. In order to provide a better deal for all stakeholders, including the butchery, consumers, and marginalized pastoralists, the Livestock Stakeholders Self-help Association (LISSA) was founded.

ii. The Kisii Network for Ecological Agricultural Development (KNEAD) assisted the villages in creating their own water resource management teams, creating their own plans for the future, providing the resources necessary to carry out the plans, and only requesting from the project the resources they actually require.

iii. Through improvements in cooperative leadership, more inclusive decision-making procedures, and higher pricing to farmers, the Gitundu-Gataro farmers group of the Nyeri Ecological Farmers Association (NEFA) helped their communities avoid violence and catastrophic splits of coffee cooperatives into smaller units.

iv. In order to ensure that dried-up streams have enough flow to water their vegetable gardens, the Nyeri Ecological Farmers Association (NEFA) Mugumoini farmers group assisted their community in stopping logging and burning for charcoal and beginning to plant trees in the forest, around springs, and along streams.

v. The Itemeni farmers' organization Nyeri Ecological Farmers Association (NEFA) assisted (a) parents and teachers in getting rid of a corrupt school board at the local primary school; (b) dairy farmers, transport operators, and "ecological" farmers dealt with the waste left behind by an outbreak of cholera caused by a rapid increase in "backyard" livestock; and (c) villagers and council members avoided conflict over the proposed leasing of the community's forest to a tea factory by having the hill gazette as a protected area.

2. Tanzania

i. In Vidunda, Msimba, and Kisanga in the Kilosa District, the Multi-sector Learning Coalition (TMLC) village multi-stakeholder "teams" pushed their local authorities to enact by-laws prohibiting cultivation close to water sources and burning forests to clear land. Through collaboration and alliances with extension, organizations were able to improve irrigation systems by creating furrows, terracing lower valley slopes, and planting trees around depleted spring lines. To get a new road to the village in Vidunda, the team organized the locals and forged alliances with a range of service providers.

ii. In collaboration with extension and another project, the Multi-sector Learning Coalition (TMLC) teams in Lushoto District's Malindi Ward organized their community to terrace sloping fields. Women's groups with goats and men's groups with cattle were able to establish zero-grazing dairy businesses with the support of the Tanga Dairy Trust. The group organized neighborhood workers and pushed district officials to provide trucks and graders to fix two kilometers of deteriorated road.

3. Uganda

i. Through the introduction of gardening technology, improved methods for communicating grievances, and demarcation of dwelling areas, Development Support Services (DSS) consulting work with UNHCR's refugee camps reduced conflict within refugee households as well as between tribal groupings.

ii. With DSS help, Veterinaries Sans Frontières (VSF) was able to provide livestock owners with better services, particularly with regard to the location and upkeep of water sources. Due to its effectiveness, VSF was able to get more money to expand its program to include two more districts.

iii. With assistance from the DSS, Self-Help Development International (SHDI) created a participatory monitoring and evaluation system that takes into account farmers' future goals, the activities necessary to realize those goals, and the creation of a framework for tracking and evaluating those efforts.

7.3.3. Weaknesses

The catalytic action's goal was to start the process of farmer empowerment so that the "demand driven" extension could proceed according to schedule. Although this has happened among people taking part in the catalytic acts, there hasn't been much of a shift in how district or national extension offices are generally administered. When it comes to public goods like roads and water supply, farmer demands for services to better natural resource management have "clashed" with "top-down" interests in commodities and company development, where national authorities have taken heed. The catalytic acts have not extended to other districts, despite the fact that farmer demand has surfaced among those directly involved. Actually, the only areas it has spread to are a village's nearby neighbors. Although invitations have been sent out by neighboring villages and districts, the resources of the local population are insufficient to accept them. With the exception of Danida's assistance in Tanzania, the multi-stakeholder organizations in Kenya, Uganda, and Tanzania have all been unable to secure finance for their proposals to scale up this catalytic action.

7.3.4. Strengths and Success Factors

Success of this catalytic initiative is due largely to the following factors:

- Ensuring that local stakeholders take ownership of collaborative learning processes by empowering them to self-organize at the village and district levels and by using organizations with convening power and legitimacy to bring various stakeholders together.

- Starting the catalytic action as a self-help endeavor outside of traditional project frameworks to prevent "hijacking" by individuals driven by projects and money. When there are no money rewards, only those who are dedicated, earnest, and perceive a return on their labor participate.
- Employing vision-based planning, which gives farmers the authority to choose how they want to develop in the future and what services they need to meet those goals. Community members are encouraged to work together when short-term problem solving gives way to the realization of longer-term objectives.
- Promoting cooperative actions and mindsets so that all parties involved can speak honestly with one another, rely on one another's honesty, and avoid having any hidden intentions. The actions in a learning process are never separated from the attitudes and behavior that contribute to its success.
- Encouraging direct communication amongst stakeholders ensures that their perspectives and interests are taken into account, assisting them in realizing their own interdependencies and the benefits that come from working together. Furthermore, peer pressure forces individuals to take personal responsibility for their past deeds and future choices.

7.3.5. Sustainability and Replicability

Farmers' ability to voice demands is likely to be sustained because this catalytic action was carried out outside of a "project" framework and relied on self-help among all local stakeholders. Facilitating multi-stakeholder collaborative learning seems sustainable at the farmer level because the advantages of resolving disagreements and crises outweigh the operating expenses of the process. But this procedure is only applied when necessary. Because a multi-stakeholder "team" at the village level is run entirely by volunteers, it seems viable. Nonetheless, the team's performance and the leadership of the community determine survival. At the district level, a multi-stakeholder group seems sustainable only if the organizations involved believe that it is assisting their employees in "doing their jobs better." District organizations don't seem to do much at all. To replicate this catalytic effect across the country, the following would be needed:

- Introducing mass media information campaigns;
- Upgrading of communications systems linking experienced villagers with interested villages for peer-to-peer training;
- Promoting vision-based planning at the village by local planning officers;
- Developing a pool of mentors at the district level to support village peer-to-peer training and district multi-stakeholder collaborative learning workshops; and
- Establishing village bank accounts to pay for mentoring services and peer-to-peer exchanges.

7.3.6. Lessons Learned

The key takeaways from the past four years have been to challenge the barriers that prevent farmers from voicing their needs and to alter public perceptions. Changing people's thought processes entails:

- Farmers should stop focusing on the issues of the present and begin considering how their community should develop in the future. They should also shift from basing community plans on the challenges of the present to the goals of the future.
- It is recommended that project managers relinquish control over project resources and responsibilities to local stakeholders and instead assign part of them to them. transferring the "economic stick" to farmers in order to get the support and supplies they require.
- The government and donors should cease requiring the standard project outcomes and instead begin to value the process as well as behavioral and attitudinal changes. They should also cease sponsoring activities and instead begin allocating funds based on the impact and results of the project. To effectively address the risks to farmer demand, one must address.
- Organizations lacking in dynamism are those that are reluctant to abandon inefficient procedures and offer their employees novel opportunities. In order to satisfy farmers' aspirations, service providers will need to abandon inefficient methods and implement new ones.
- Leaders of farmer organizations, service providers, and the government fail to acknowledge the importance of time invested

in multi-stakeholder coalitions for the continued existence of their organizations.

- Managers and politicians' lack of faith in farmers' ability to innovate and organize themselves. Authorities, including officials, view, and consider farmers as obstinate and illiterate, lending little weight to their expertise and experience.
- farmers' lack of control over financial resources and operational decisions. In order to compensate service providers, farmers should be in charge of their financial resources and create their own operational policies for demand-driven extension.

Demand-driven extension systems have been installed with the help of a variety of techniques. Promoting self-organizations at the district and village levels that necessitate capacity building, as well as investigating interest in multi-stakeholder collaborative learning with a variety of stakeholders, are two ways to start such programs. Sensitization workshops for national/district level organizations, multi-stakeholder collaborative learning workshops at the district level, and self-organized multi-stakeholder groups at the district and village level can all serve as a basis for directing demand-driven extension initiatives. Lastly, tools that help keep groups going for demand-driven extension include: links to global experiences through the internet and outside mentors; follow-up "reflection" visits by mentors and email exchanges; and the sharing of success stories amongst farmer groups facilitated by mentors.

7.4. FISCAL SUSTAINABILITY OF THE FARMER FIELD SCHOOL APPROACH

Agricultural extension is a valuable public utility that should be supported by the public in many developing nations. But for large-scale agricultural extension systems in both wealthy and developing nations (Feder et al., 1999; Hanson and Just, 2001), fiscal sustainability has proven to be a common issue. The financial burden that expensive national systems imposed on public budgets was unsustainable, which led to their drastic reduction or elimination. Purcell and Anderson (1997) reviewed agricultural extension projects supported by the World Bank between 1977 and 1992 and discovered that a significant percentage of free-standing projects (76%) had an uncertain or unlikely sustainability rating, indicating that "inadequate funds to operate the services properly was a common phenomenon." The majority of borrowers faced significant challenges in covering the ongoing expenses of national extension services, to the point where in numerous instances the viability of the established systems was questioned, they continued. This case study explores how the farmer field school (FFS) strategy, a national agricultural extension tool, was used in the Philippines and Indonesia to address the issue of fiscal sustainability. The ability to continue the extension effort at a level that may reasonably be expected to achieve substantial coverage (directly or indirectly) of the farming population across the country is what we refer to as fiscal sustainability.

7.4.1. The Farmer Field School Extension Approach

Originally intended to educate irrigated rice farmers in Asia on integrated pest management (IPM), the Farmer Field School Extension (FFS) strategy was created. Indonesia and the Philippines played a major role in carrying out this extension operation. Since then, IPM-FFS experiences in these two nations have been chronicled and utilized to support and extend FFS and FFS-like initiatives to other nations and crops. Many developing nations are currently implementing FFS activities, albeit very few run FFS as a national system. As of right now, a typical FFS covers the practical aspects of "plant health, water management, weather, weed density, disease surveillance, plus observation and collection of insect pests and beneficials" and instructs farmer participants in agro-ecosystems analysis, also known as integrated pest and crop management (IPCM) (Indonesian National IPM Program Secretariat 1991). In order to transform field school participants into "confident pest experts, self-teaching experimenters, and effective trainers of other farmers," the FFS strategy depends on participatory training techniques (Wiebers, 1993). During a single crop-growing season, an archetypal FFS involves eight to 12 half-day sessions of practical farmer experimentation and non-formal training for a group of twenty to twenty-five farmers. This village-level program is initially run by paid trainers who concentrate on providing components and workable solutions for general good crop management practices, along with problem-solving techniques for pest management. Participants hone their decision-making skills through group interactions, and they gain empowerment by mastering management, leadership, and communication techniques (Van de Fliert, 1993).

7.4.2. Farmer Field Schools and Fiscal Sustainability

The financial outlay required to maintain such an endeavor, especially at the national level, is a significant drawback to endorsing FFS as an agricultural extension strategy. A substantial financial commitment is suggested if government intends to implement a large-scale training program over an extended period of time using official trainers, which may not be financially feasible. The farmer-trainers concept has been one way to lessen the financial load. The idea is to lessen FFS's reliance on government funding support by encouraging graduates of the program to enroll in training of farmer-trainers (TOFT) and then have them educate additional farmers. The growth and success of farmer-led schools will determine how long the field-school model lasts. Farmer-to-farmer field school teaching is seen as a promising means of increasing FFS coverage. It is important to take into account the possibility of information dissemination among farmers through informal contact, in addition to farmer-led schools, which can be thought of as a formal diffusion mechanism. But since the goal of the training is "to help farmers develop their analytical skills, critical thinking, and creativity, and to help them learn to make better decisions," it should be highlighted that the content delivered during an FFS is very difficult (Kenmore, 1998). "Farmers master a process of learning that can be applied continuously, rather than a specific set of contents of messages." Because of this, the FFS curriculum covers intricate ideas related to agroecosystems and decision-making processes that are unlikely to be of daily interest to farmers if discussed informally through oral communications. This case study contends that formal farmer-led school coverage is limited, informal exchange has little impact, and official financing is heavily relied upon in the Philippines and Indonesia, the countries with the longest experiences with FFS as an extension approach. As a result, the sustainability issue remains unresolved.

7.4.3. The Philippines Case

National IPM-FFS initiatives in the Philippines, officially dubbed the KASAKALIKASAN program, were initially launched as a Department of Agriculture five-year program (1993–1997). As of right now, the program's primary funding source is still the federal government. The annual budget of the central government sets the financing levels. The business sector, certain NGOs, and municipal governments provide the meagre program resources. According to a SEAMEO Regional Centre for Graduate Study and Research in Agriculture (SEARCA, 1997) evaluation of KASAKALIKASAN, the program trained 183,829 farmers in 7202 farmer field schools between 1993 and 1997. According to the community IPM program's webpage (http://communityipm.org/philippines), around 200,000 farmers have received training by October 2000. For the five-year period from 1993 to 1997, PhP 235 million (US$8.75 million) was the program's budget, or PhP 1280 (US$47.6) each trained farmer. Local government entities only contributed 9.4% of this total; the Department of Agriculture and other central government agencies provided 87.3% of the funding. It would take more than 15 years and cost over US$47.6 million if public resources were to be continuously relied upon at present spending levels in order to have one million Filipino farmers attend at least one FFS. This represents 20% of the country's estimated 5.0 million farm households. The likelihood of field schools covering a sizable portion of the farming community through meaningful means is depressing given this poor pace. It's also critical to remember that the real expenses will probably be greater than those listed in the SEARCA 1997 assessment. Only direct program appropriations—that is, expenses incurred and funded by program funds—are included in the reported budget projections. They exclude indirect costs, such as those incurred for central, provincial, and local government administrative and other human resource costs that are not directly covered by program monies that have been allotted. As previously stated, the FFS method relies on the dissemination of knowledge and skills gained through FFS through informal farmer channels or more formal farmer-led FFS activities in order to be a sustainable national agricultural extension program. On the other hand, the actual data from the Philippines suggests that these diffusion pathways are not important. First, a recent study conducted in the Philippines by Rola et al. (2002) found that there were no significant variations in knowledge ratings between "FFS –knowledge recipients" and other non-FFS farmers, but there were significant disparities between FFS grads and non-graduates. According to Rola et al. (2002), there isn't much knowledge that FFS graduates impart to other farmers. The reason this is maybe not too shocking is explained by a previous study by Rola et al. (1998). This previous study reports that "FFS graduates mentioned their willingness to share their notes, although it was unclear whether they were willing to spend time in teaching in the field" during the focus group talks that were held. Second, the Philippines FFS program has not placed much emphasis on farmer TOFT graduates. Only six TOFT sessions under KASAKALIKASAN were held between 1994 and 1997, with roughly 25 selected FFS graduates present every session, according to the SEARCA (1997) study. In this situation, the FFS technique in the Philippines will only achieve extremely limited coverage unless there is a concerted effort at farmer-to-farmer knowledge transfer and state or NGO backing of follow-up operations. The continuation of extensive official engagement in the absence of such funding brings up the issue of fiscal sustainability, which has plagued big extension networks for a long time.

7.4.4. The Indonesian Case

IPM-focused farmer field schools in Indonesia were first funded (from 1989) by the FAO and USAID, and then (since 1994) by the World Bank with additional funding from USAID. The Indonesian national program, more than that of the Philippines, strives to spread IPM and other information by supporting farmer-led and farmer-initiated FFS initiatives. It was anticipated that skilled farmer-trainers would take center stage in planning and executing FFSs (Braun, 1997). Moreover, there will be less financial burden associated with publicly-funded field school investments if some of the costs of farmer-led FFSs, such as participant compensation, food expenses, experimental plot rent, and trainer honoraria, are eventually covered by local communities. There are differing opinions regarding whether the IPM Training Project will be viable after the World Bank's 1999 loan to it ended. There are questions concerning the project's sustainability because of the World Bank's internal operations appraisal process. However, some studies contend that networks of farmer-trainers, who provide the majority of training in Indonesia, have already supplanted professional trainer-led FFSs. To evaluate the validity of these statements, we look at empirical data on the type and prevalence of FFS training from two distinct farm-level surveys.

7.5. SEMI-PRIVATIZED EXTENSION CIRCLES

In Germany, the federal states are in charge of agricultural extension (Hoffmann et al., 2000). Therefore, due to historical factors and regional specificities, a range of systems for organizing and supporting agricultural extension have arisen notwithstanding the common framework of EU and federal agricultural policy (Hoffmann, 2000). Nonetheless, pressure from budget cuts and the global conversation about how to provide agricultural services more efficiently resulted in adjustments in every state. These state-specific modifications are currently being implemented.

7.5.1. Extension Services in the Federal State of Baden - Württemberg

About 86,000 farmers labor the countryside of Baden-Württemberg, a southern German state with an area of 37,000 km² and a population of about 10.5 million, on their small and medium-sized farms (average size: 19.4 ha.). It was determined to create a public extension system following World War II to help farmers increase productivity and output in order to guarantee food security at affordable rates. This approach was validated in 1972 by a state law on agriculture that mandated that agricultural extension be provided to recipients at no expense to them (MELF, 1972). Given the wide range of production systems and economic standards among farm families and businesses, the idea was to ensure that all of them would have access to a sufficient extension service. Extension was thus provided by regional branch offices of the state ministry of agriculture, and the extensionists were state workers. A multitude of issues propelled agricultural extension change. Although the 1972 law is still in effect, during the 1970s and 1980s, agricultural extension circumstances underwent significant changes. Basic food security was no longer a problem during the EU's overproduction period and could not be used as an excuse for government expenditures to promote increased agricultural productivity and output. In public discourse, there has been a growing questioning of public agricultural extension. Public attention has shifted away from productivity and towards the preservation of the environment and natural landscape as well as the sustainability of agricultural production systems. Extension faced the issue of realigning its tactics and material appropriately. Pressure on agricultural commodity prices increased, driving many farmers out of business. Between 1955 and 1995, there was a 70% decrease in the number of farms. Numerous more farms were facing dire circumstances. As a result, there was a surge in demand for highly focused advising services pertaining to basic life planning decisions and future farming prospects. Rather than providing a wide variety of items as a risk-minimizing strategy, the remaining farm production was increasingly concentrated in specialized units focusing on certain production lines (i.e., dairy, fattening). Farmers began to seek more detailed and specialized production information and assistance as a result of this. The current extension system was unable to adequately address the challenge posed by the growing need for more specialized services and other changing demands, in addition to juggling additional administrative responsibilities. The staff of regional branch offices almost drowned in the 1980s due to EU laws and programs, as well as the establishment of income transfers for landscape maintenance. The amount of staff time allotted to developing and providing technical consulting services was decreasing with time. The combination of these variables created a situation in the late 1980s that both consultants and farmers found frustrating. It was time for reform (Grosskopf, 1989). In order to refocus advisory efforts on newly emerging core tasks, such as providing advice for sustainable and environmentally-friendly production and animal rearing in accordance with moral guidelines for animal welfare, and developing and implementing farm business plans required to address the challenges of individual farms and households, public extension system staff members needed to be relieved of their excessive workloads. This meant that comprehensive information and guidance on particular production and economic issues faced by specialty farms have to be supplied without resulting in significant additional public costs.

7.5.2. Extension Circles as the Heart of Reform

A reform paper (MLRELF, 1989) on the advancement of agricultural extension and administration was released by the state Ministry of Agriculture (MOA) in 1989. A semi-privatized extension structure based on so-called "extension circles" was proposed in this study. The reform's main goals were to: (a) address the expanding need for advice on particular production and farm economic issues faced by expanding, specialized farms (MLRELF, 1991); and (b) lessen the workload of public extension officers by assisting them in refocusing their efforts on the recently developed area of publicly financed advice. A group of farmers who formally associate and elect a board made up of farmer members and one employee from the MOA's regional administration branch jointly establish an extension circle. Up to five advisers may be appointed by the board, with roughly 50 farmers for every adviser. The ministry receives an application from the association for recognition and support. Following recognition by the MoA, the association is qualified to receive MoA support for:

- 50% of personal and technical costs (up to 28,500 Euro per adviser),
- Technical and methodological training of the adviser,
- Use of infrastructure of the regional MOA branch offices, and
- Participation in the technical research and information system of the MOA (MLRELF, 1998).

7.5.3. Development of the Model

With the full cooperation of MOA branch offices, 21 model extension circles were established in 1989 (MLRELF, 1991). Three of them focused on ecological agriculture, nine on intensive gardening and horticulture, and nine on crop and animal production. Now, there are roughly 50 recognized associations with 82 advisers working for them. Even so, this figure still does not account for all farms in Baden-Württemberg. Approximately 10% of farmers are currently covered by the program, with a 50:1 farmer-adviser ratio. With agriculture's increasing structural changes and specialization, there is undoubtedly a greater potential need for services under such a strategy. There could be a number of reasons why associations are not growing more quickly. The priority appears to be the following three reasons:

- The ministry came up with the concept for the advisory association and carried it out. The ministry also aggressively encouraged the regional branch offices to assist the formation of the first 21 groups. This is no longer the case, as numerous associations founded by farmers already exist. Even yet, getting MOA assistance still requires a lot of work in terms of planning and completing the paperwork. For diligent farmers who are frequently highly focused on managing their own farm, this presents a challenge.
- The advisors' and circles' actions are solely restricted to providing information and knowledge support for issues related to

agricultural economics and productivity. State funding would be withheld if collaborative marketing initiatives or input provision were organized under an advisory association (MLRELF, 1997). The MOA is adamant about this regulation because it prevents rural trade and service businesses from being less competitive. Because of this restriction, prospective members are discouraged from joining circles for fear that the 750 Euro yearly membership fees may not be worth the increased value of extension services.

- Staff members at the MOA regional branch were extremely guarded and dubious in the early years following the model's implementation because they were worried about competition and skill erosion. The inability to clearly define distinct tasks and distinguish between the circles' and the public extension's tasks continues to foster this mindset.23 Consequently, advisors in the public sector have not fully supported the establishment of new advisory associations. Additionally, there has been occasional conflict when pre-existing associations coexist with MOA regional chapters, sharing offices and infrastructure.

The goal of MOA's 1989 reform was to provide a closely related, semi-privatized production advice service to supplement the regional branch offices. Generally speaking, no circle hired more than one adviser from the regional branch office. This idea has lost some of its luster in recent years as associations grow more autonomous. A number of circles have merged in the past few years in an effort to strengthen their own infrastructure and organizational structure while overcoming the relative isolation of their advisors.

7.5.4. Impacts of the Reform

Examining a reform's effects typically centers on the following query: Did reform initiatives solve or at least decrease the problems they addressed? Investigating a little more thoroughly could reveal effects—positive or negative—that were not initially anticipated. In this instance, the first two queries would be: (a) Could specialized farms benefit from the extension circles' assistance in resolving their production and financial issues? And (b) Did the creation of extension circles reduce public extension officers' workloads and allow them to refocus on extensions? Regarding the second category of impacts—those that the reform did not specifically target—two concerns come to light. The first concerns the working relationship between extensionists and farmers as well as the quality of services provided. The implications for the state budget are the subject of the second concern. Circle Members' Satisfaction and the

Improvement of On-Farm Results When asked to rate the effects of advice on their own farms, 80% of Baden-Württemberg extension circle members gave high positive evaluations (acceptable to excellent), according to a recent survey (Gruber, 2002). Consistent circular expansion also suggests satisfaction with the model. The increase in the number of members in the current circles appears to be even more persuasive in this regard. There has been a 250% rise in membership at three of the original associations. An additional measure of the value of the circles' extension services may be a comparison of the output outcomes.

The production data for one association show highly favorable outcomes of the advising effort, despite the fact that attribution of impact is often challenging (was it actually the advisory support or were changed framing conditions responsible for the effects?). After participating in this advice circle for a few years, farmers typically see improvements in productivity and farm management abilities. One of the first circles' 10-year results comparison shows that member farms increased their land by 56% and their milking cow population by 29% on average. Productivity increased by 27% as indicated by the annual milk output per cow (Willige, 1999).

7.5.5. Facilitation of Public Services Reorientation

Whether public administration and extension officers could realign their activities and accomplish their remaining tasks more efficiently is the second question. The response has two parts. The demand on public extensionists to always be up to date on all recent production-related technical breakthroughs was definitely relieved with the advent of specialized production-related advice.

While some public extensionists lament their lack of technical prowess, this allowed them to focus more on other assignments. The extra time that public sector extension employees had on their hands was typically taken up by more administrative work, particularly enforcing EU regulations and providing guidance on how to set up production and complete forms in order to apply for transfer payments from the state, the federal government, or the EU. These transfer payments, which can account for up to 30% of average farm revenue, are typically linked to environmentally sound production guidelines or landscape protection measures.

In that regard, there was unquestionably a shift towards public extension that contained more public goods. But this didn't happen because the branch offices or the administration as a whole didn't go through a conscious reorientation process that would have identified expanded working capacity and concentrated it on the newly emerging demands.

7.5.6. Impacts on the Public Budget

With an annual payment of up to 50% of the entire expenses or a maximum of 28,500 Euro per adviser, the state provides help to the associations. Every year, following their initial registration and approval by the MOA, associations are required to submit an application for funding. The 28,500 Euro ceiling has not increased over the past few years, meaning that the state contribution is currently no more than 40% for a number of circles. The association members' portion of the remaining expenses is typically distributed based on the following idea: Each member farm contributes based on its production resources (e.g., arable land, milking cows, and fattening areas) in addition to an annual baseline sum. The savings for the state budget appear to be significant when one considers that staff costs for public employees can reach up to 50,000 euros, and additional running costs for an adviser can reach an additional 20,000 euros. However, real cost savings shouldn't always be anticipated, given that the change was a response to an extremely inadequate delivery scenario that demanded adjustment. It is true that there are no budgetary savings for extensions when comparing the pre- and post-advisory circle scenarios. In the near term, however, the situation is really rather reversed: budget increases resulted from financial assistance for the advisory circles since the agricultural administration and extension branches were too heavily burdened with administrative duties to be reduced. Over the past 10 years, there has been a decline in the number of administrative staff members; nevertheless, the presence of extension circles is not a primary cause of this reduction. It is necessary to ask the opposite question when discussing budget savings: Would it have been less expensive to expand the public extension system in order to meet the growing demand for advisory services in terms of both quantity and quality? Without a doubt, the response is no.

7.5.7. Quality of Advisory Services and Working Relationships

The associations' members now receive much better advice services in terms of quality. Each member can anticipate at least four days of direct contact with the adviser in a 50:1 farmer adviser ratio (equating to 200 working days annually). In contrast to public advising services, which offer farmers personalized, one-time assistance upon request, association advisers offer their farmers ongoing help through activities including annual profitability assessments, collective internal benchmarking, experience sharing, and newsletter distribution. The old role-conflict issue is history when an adviser is relieved of administrative duties and responsibility for enforcing laws and regulations; the likelihood of a closer, more trustworthy relationship between adviser and farmer increases accordingly. However, the adviser has certain issues with this working relationship. Due to the fact that their clients are also their employers, advisers may face high service demands and even become targets of exploitation. Circle advisers typically put in a lot of work and have limited free time (Willige, 1999). It can be difficult for them to survive at times due to this and the fact that they frequently feel like "a lonely fighter" with no prospects for advancement inside the association. After less than two years, almost 30% of the employees quit because they are dissatisfied or have had arguments with other circle members. After around seven years, others frequently give up, trading in their expertise and practical brilliance for more optimistic outlooks. The member farmers specify what they need and what areas the consulting work will concentrate on. The association's membership is highly satisfied as a result of this. Experience, however, demonstrates that competent association leadership is required to get all members to agree on operational goals for a certain time frame and to convert these goals into terms of reference that an adviser can work with. If this doesn't occur, there will be a difficulty with consistency in understanding of roles and priorities, and the consultant runs the risk of being overworked or lost when faced with a different notion on each farm. Durability of Change The main questions about the sustainability of the changes are whether the forces that initially prompted them will continue, and whether the stakeholders will eventually find the reform's outcomes acceptable. First, it is obvious that the trend towards middle-sized farms with a concentration on specialized production will continue, and there will undoubtedly be a rise in demand for specialized production skills. Furthermore, the reorientation of public advisory services towards agro-environmental and more comprehensive themes like farm or even rural development is unlikely to reverse, meaning that administrative demands on the public service will not decrease (Koch, 2001). We most likely get into the issue of costs and financial burden sharing while discussing the reform's long-term viability. The state continues to have a strong interest in rural areas and the farming industry. Therefore, co-financing of the associations shouldn't be a major issue, especially considering the relatively small annual burden of about three million euros. There is some uncertainty, nevertheless, because state co-finance is reliant on decisions about the biannual budget. A decrease in the state share can be taken into consideration if the number of co-financing applications increases significantly. Infrastructure, data, and—above all—training for advisers will undoubtedly be made available, provided that doing so is feasible given the constraints of the public system's current organizational structure. The growing different training requirements of highly specialized circle advisers and generalist public advisers could present an impending challenge in this regard. It is unlikely that the state will design and carry out an additional training course for the association advisers. Conversely, farmers who participate in the circles accept and are growing accustomed to paying a portion of the fees for guidance.

However, past experiences indicate that middle-sized farms will only contribute to extension delivery expenses up to a certain level (Currle and Schutz, 2001; Nagel et al, 2002). Farmers typically leave organizations in large numbers if state payments drop below 30%. In conclusion, the reform's viability appears assured, provided that significant adjustments are not made to the state and farmer's financial burden-sharing arrangement, which is the most delicate area.

7.5.8. Conditions for Scaling-up and Transferring the Approach

Replicating and scaling up reform components globally would need taking into account a number of prerequisites. The strategy performs well in a setting of medium-sized, highly integrated farms with specialized crops. In a subsistence setting, it most likely wouldn't be feasible due to the extreme lack of money. Applying the group strategy to huge industrialized farms that are geographically apart from one another and possibly competitors in the same market if they specialize in the same producing area would also be challenging. If the funding issue can be fixed, extension circles appear to be a better option for small farmers than government extension services. This might be made feasible by agreements in which farmers pay only 10% of the total cost of services and receive a 90% subsidy from sponsors in the government or nonprofit sector. This strategy will undoubtedly perform better in a setting where farmers who grow crops for their families' subsistence also cultivate a cash crop that generates some revenue.

7.5.9. Lessons Learned

7.5.9.1. Allow and Facilitate Steady Adaptation

The example given provides a generally effective model for addressing some of the most significant challenges associated with public extension. A 12 year track record demonstrates that state-subsidized extension associations are capable of offering farmers targeted, ongoing assistance with production-related issues and farm economics. For middle-sized farms who cannot afford to pay the full fees of a private consultant, this model offers efficient extension services (i.e., an acceptable client-extension ratio, well-trained experts). Two key conclusions can be made from a closer examination of the conflicts that existed between state administration employees and the new institutions that dominated the program's initial phase:

- In addition to being crucial at the central level of administration, the political will to carry out the change must be transparently communicated to the level of daily cooperation (Rivera and Zijp, 2002).

- A clear redefining of tasks and early involvement of lower level concerned actors are part of this transparent communication approach. A deliberate approach to this job reorientation aids in identifying and optimizing the extra capacity that has been freed up in public administration.

Program guidelines and policies should be flexible because the agricultural industry and administrative landscape are changing quickly. Advisory circles should have more autonomy as associations and advisers develop their skills and as the number of members of extension circles rises. In the early stages of a circle's development, very close administrative support was crucial, but as the circles have grown, this support is no longer required.

7.5.9.2. Train Farmers to Manage a Service Enterprise

One challenge that undoubtedly arises with the introduction of task-oriented farmer groups is the occasionally inadequate ability of group members to lead that group, to oversee and assist decision-making, and to adequately direct one or more staff (advisers). Public administration is required to offer support and training in this regard. Public extension services are facing a new difficulty with training in these talents, and future responsibilities will probably center around this. However, only recently established associations should have direct public administration engagement in circle management during their startup phase.

7.5.9.3. Prepare Advisers Adequately

The circles' advisers need to be prepared for their work, but the training they receive at college or university is unquestionably insufficient. The farm administration should establish an efficient introduction training program and support it with frequent meetings where experience can be shared. This would combat the "lonely-fighter-problems" that exist in some of the rings and promote unity and togetherness among the advisers. It is crucial to prevent huge early dropouts after less than two years, even though it is perfectly normal for young advisers to depart after about five to seven years in order to pursue new professional challenges inaccessible in an association with at most five personnel.

7.5.9.4. It Is Not a Cheap Solution

As previously mentioned, the approach of providing subsidized extension advisers is very focused and customized to address the daily requirements of participating farmers. The extensive assistance provided to particular farms necessitates a high adviser-to-client ratio. In Baden-Württemberg, one public adviser (300:1) was replaced with six circle advisers (50:1). The cost per adviser won't decrease six-fold, even with a sizable contribution from the client-farmers. Three choices remain in this situation:

- The state can agree that additional financing is needed to provide farmers with better advice services.
- It is possible to stretch the adviser-client ratio towards the current numbers for public extension by reducing the number of advisers for circle extensions. This approach, however, carries the danger that farmers won't feel that their income and output are being significantly increased, and as a result, they won't be prepared to pay their membership dues.
- In addition, farmers serviced by a subsidized extension circle program can be carefully targeted to address unique requirements and not exceed a set percentage of the total farmer population. The public advisory service can be kept in place.

7.6. PORTUGAL EXTENSION REFORM IN THE INTERIOR NORTH

Portugal has implemented extensions in four main phases over the last 30 years:

- A period of the late 1970s and early 1980s when public extension activities were actively developed; during this time, national, regional, and local services were organized, extension agents were trained, and programs and actions were created.
- A period of decline in public activity that began in the middle to end of the 1980s and was marked by national extension campaigns like "100 Days in the Field" and "Producing Better"; during this time, agents were also expected to handle administrative duties, which kept them away from the day-to-day issues that farmers and rural communities faced; public services were also bureaucratized, and field contact gradually decreased.
- A phase of privatization that began in the early 1990s with PROAGRI, a government initiative designed to improve farmers' organizations' capacity for management and provision of technical support to both members and non-members.
- A phase of diversified extension service delivery, coinciding with the current circumstances, wherein private businesses, farmer associations, cooperatives, and other organizations carry out

the extension function in a relatively dispersed and fragmented manner, with the exception of certain networks and coordination. Public extension is essentially nonexistent in this scenario. The early 1990s reforms and modifications are examined in this case study.

"In general, and in practical terms, extension in Portugal remained in the hands of isolated technicians, lacking institutional support (from the point of view of monitoring and evaluation, as well as for research back up), with the obligation of developing concurrent tasks, in an institutional environment marked, among other things, by the lack of professional motivations, and, many times, by uncertainty regarding the agricultural policy measures," according to a review of Portuguese extension experience written in the early 1990s (Portela e Cristóvão, 1991:49). Furthermore, a thorough examination of rural development initiatives in Portugal's interior north showed that the three main tenets of extension practice were as follows:

- A "top-down" view, providing little room for farmers' participation,
- A selective and elitist approach, which underestimated social and farming diversity and concentrated attention on a small segment of the population, and
- A narrow or linear perspective of extension-research-training processes (Cristóvão e Portela, 1995).

7.6.1. Description of the Reform Measures

PROAGRI, a program element of a larger scheme created following Portugal's entry into the EEC and targeted at the growth of Portuguese agriculture, marked the beginning of reform efforts in 1989. PROAGRI's main goals were to improve farmer organizations' ability to intervene in the provision of extension services to their associates and to farmers in general, as well as to strengthen the technical and managerial capacities of farmer organizations. Essentially, the fundamental concept involved privatizing extension, that is shifting this duty from the State to farmer associations, in line with the widespread European tendency of involving farmers in the execution of technical assistance and agricultural development (MAPA-DGPA, 1989). Three significant farmer organizations were consulted in the development of this reform by the Ministry of Agriculture:

- The Confederation of Agricultural Cooperatives (CONFAGRI).
- The Confederation of Portuguese Farmers (CAP).

- The Portuguese Association of Young Farmers (AJAP).

Although PROAGRI was originally intended to run until 1993, it was superseded by other programs with comparable goals, first under PAMAF (1994–1999) and then under AGRIS (2000–2006), which is currently restricted to forestry cooperatives and associations. The initiative was primarily aimed at farmer organizations, such as forestry associations, vertical, and horizontal associations, agricultural, and forestry cooperatives, and other groups approved by the Ministry of Agriculture. All organizations were welcome to apply, but before being granted permission to take part in the program, their associative, financial, and economic viability were evaluated. The groups had to provide a plan for development and investment. It is possible to receive funding from the State and European Communities:

- To strengthen the organization, and
- To support the creation of extension services (e.g., creation, and development of extension capacities; improvement of extension activities; initial and in-service training of extension agents).

The organizations were in charge of co-financing the suggested activities, with each activity area having a varied level of state and EU assistance and a maximum of 80–90% in the first year of implementation and 50–60% in the fifth. Support from the state and the EU was greatest for businesses that operated in underprivileged areas. Organizations that violate program objectives or improperly exploit financial resources allocated to them are subject to sanctions.

7.6.2. Impacts of Reform in the Interior North

Over the previous 10 to 12 years, the revised system was developed. The subsequent numerical components offer an outlook on the evolution of reform in the Trás-os-Montes region, located in the interior north of Portugal. About 400 technicians/extension agents work for about 150 farmer organizations (54 associations, 40 cooperatives, 24 wine cooperatives, and 30 management centers). The majority of these agents were born in Trás-os-Montes into families involved in agriculture (about 50%); most have a higher education degree in agriculture or a related field that they obtained in the region approximately 6–7 years ago; most are relatively young (average age of 31 years); and most regularly attend continuing education courses (average of one per year). Depending on the kind of organization, there are different numbers of agents: in wine cooperatives and associations, there are three; in management centers and other cooperatives, there are around two. Based on the technical and extension staff of farmer organizations, the new agricultural knowledge and information system supporting agricultural growth in Trás-os-Montes is an offshoot of PROAGRI and its successor programs. Extension agents, also known as farmer organization technicians, play a crucial role in facilitating the flow of information and knowledge, primarily of three categories:

- Legal-bureaucratic,
- Technical or production related, and
- Social.

The first two categories, legal-bureaucratic, and production-related, are the outcome of the intricate framework that currently governs agriculture in the member states of the European Union. This framework consists of production standards, farm support systems, environmental, and sanitary regulations, fiscal laws, and various support initiatives aimed at encouraging the modernization of production systems and enhancing the qualifications of farmers. The social type pertains to the extensive range of requirements that farmers have, surpassing the legal and technical facets. This kind of knowledge is useful for fostering interpersonal bonds and mutual trust as well as for assisting in the resolution of individual and societal issues.

7.6.3. What Works Better? What Did Not Improve?

Farmers now have much easier access to information, especially about the administrative processes and regulations associated with the EU Agricultural Policy (CAP) and the agricultural law and fiscal system. This shows how successful the organizations and technicians involved are.

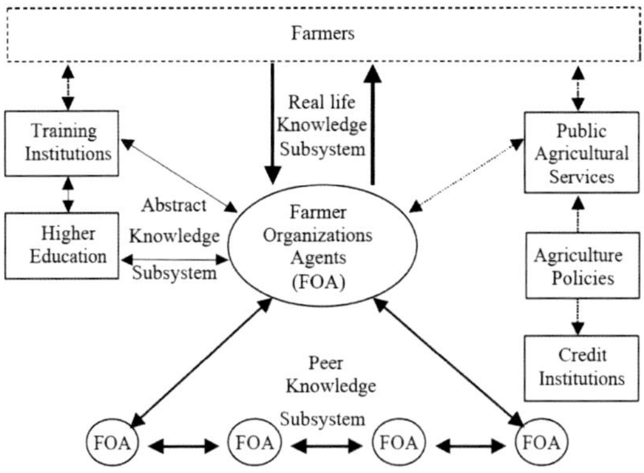

Figure 7.2. *Production system and diffusion of knowledge and information.*

Source: Alex and Rivera, (2004).

But the work that farmer organization technicians do tends to be so intensely legal and administrative that, most of the time, this consumes all of the time and resources, leaving very little or none for the technical or production related activities. This is undoubtedly an issue that needs to be addressed (Figure 7.2).

The technicians' efforts in the technical domain are further limited by the farming structure's shortcomings. However, a number of extremely important goals have been accomplished, including the introduction of accounting systems and other management practices, the planning and execution of farm investment projects, the diffusion of integrated pest management and other environmentally friendly farm practices, particularly in the cases of vineyards, fruit, and olive production, and the promotion and certification of traditional products with recognized quality (such as regional sausages, cheeses, honey, fresh, and dry fruits, and meat). It's also important to note that farmers are supported in enhancing their product marketing through more competitive means. In actuality, farmers—especially small-scale farmers—have improved market access thanks to their groups and cooperatives. However, farmer organizations confront a number of challenges, including fierce competition from commercial enterprises and member disloyalty—members who frequently sell their products to other agents and fail to adhere to internal policies. The aforementioned issue is undoubtedly a result of an emerging cooperative/associative spirit (e.g., individualistic views, lack of active engagement). A few other issues include the large number of farmer organizations operating in the areas, their high level of atomization, how hard it is to collaborate and create synergies, and how dependent they are on outside funding, which is mostly CAP-related and routed through government channels. The system's many actors have brittle links with one another, and goals and resources are unclear. Instead of through active exchanges between the primary actors, the farmer organization technicians fundamentally encourage the flows of knowledge and information by trying to create a demand for their job. For example, demand from farmers is still quite uncommon. There is virtually little interaction between farmers and farmer organization technicians and other significant system actors, such as higher education institutions, loan agencies, and state agricultural services.

7.6.4. Major Benefits

The revamped system primarily benefits farmers. Despite the issues noted, they have easier and better access to services that are necessary to ensure farming's survival and boost its competitiveness given the current institutional, political, and market environments. Actually, thc system keeps an eye on and assists farmers with their operations, assisting them in meeting official standards and promoting information and knowledge exchanges that are critical to enhancing agricultural production systems. This system's reasonably effective operation benefits the national and EU organizations in charge of agricultural development because it is the outcome of policies, programs, and financial tools from both the EU and the country. Their goals, especially involving farmers in the implementation of policies, are being met. The farmers' organizations' transfer of extension responsibilities to regional institutions, in particular, frees up more time for other, typically administrative, operations. The fact that farmer organization technicians can work and, for the most part, pursue successful professional careers makes them significant beneficiaries as well. In several cases, the professional expertise and involvement gained in organizations and cooperatives has served as a springboard for other endeavors, such the founding of private consulting firms. No research has been done on gender issues. Nonetheless, the proportion of men and women among the technicians employed by farmer organizations is comparable. However, the vast majority of farmers who lead these organizations are men. Men and women collaborate side by side to manage the farm, present development ideas, and take part in professional training exercises at the farm level.

7.6.5. Critical Success Factors

Success is largely dependent on the proximity of extension agents to farmers as well as their identification with the local people and the context for services (the distinctive features of the region's agriculture). The agents become more conscious of the need to be adaptable (in mixing various types and pieces of knowledge) and to address issues outside of the strictly technical sphere as a result of proximity and identification. The financial support provided to farmer organizations through various channels, whether direct or indirect, is another essential component of their success. This support can be in the form of funds for professional training initiatives, funding for the development of associations and cooperatives, including the acquisition of material and human resources, funding for the modernization and improvement of farms, and funding for other purposes. The contribution of the two public higher education establishments in the area—the University and the Polytechnic Institute—both of which have significant specialties in fields relating to agriculture and rural development, was another element in the success of the endeavor. These establishments offer introductory instruction, ensure chances for ongoing learning, conduct research, and distribute data. The majority of farmer organization technicians received their training from these two establishments and have a more or less ongoing, mainly informal relationship with their professors and researchers.

7.6.6. Reform Sustainability and Replicability

The promotion of agricultural methods that increase the sustainability of farming systems and the quality of the products produced is crucial to the progress of farming in the region. In addition to research, these knowledge-intensive techniques necessitate a significant contribution from efficient extension, training, and information. This primary concept emphasizes how crucial it is to preserve and improve the framework that has been developed over the past 10 years. But there is still a significant obstacle to overcome: how will the system be funded going forward, given that it has been primarily supported by financial instruments tied to CAP? There are two options that can be put forth: either the beneficiaries assume the costs, which seems more feasible for more lucrative agricultural activities like viticulture and possibly olive and chestnut production (a possible scenario, with some potential implications in terms of small farmer exclusion); or the continuation of the support levels provided by CAP (a scenario that is unlikely given the anticipated budgetary constraints). Enhancing farmers' ability to ask for and obtain high-quality services from their organizations and other institutions, enhancing the effectiveness of such organizations and making them more demand-driven, is another obstacle to sustainability. Regarding the last argument, it seems likely that a restructuring of the system is required, incorporating fewer but progressively more competent grassroots organizations. These days, it's not uncommon to see three or more cooperatives and associations in a single municipality, all of them essentially serving the same smaller "clientele." Farmers will benefit from increased education and training, as well as a growing awareness of the intricate interplay of institutional, political, legal, and market factors.

7.6.7. Lessons Learned

Lesson 1: The establishment of universities in the area was a wise strategic move that would have a lasting impact on its growth. In addition to producing competent workers, the University and the Polytechnic Institute are home to a sizable pool of specialists in the social, economic, and technological sciences. The training of extension agents and other technical employees of farmer organizations will continue to be greatly aided by these institutions. Both the promotion of in-service training possibilities (extremely focused, high-quality, short-duration courses) and the enhancement of the courses and degrees offered (especially the practical dimension better fitted to the complexity of modern times) require more attention.

Lesson 2: Achieving quality education and development takes sustained medium- to long-term work and cannot be done quickly. The University and Polytechnic Institute were established approximately 30 years ago. The farmer organization technicians received their training six to seven years ago, and it is only recently that we have been able to witness some of the outcomes of their efforts within the cooperative and association movement. Combining experience knowledge—which is gained in the field via interactions with farmers and others—with abstract knowledge—which is developed during academic preparation—takes a significant amount of time. This is a fundamental prerequisite for fostering change.

Lesson 3: It is not appropriate to abolish an outdated system without first establishing its replacement. In this instance, the public agricultural services-based system was abandoned in the late 1980s, and the redesigned system didn't begin to function satisfactorily until the late 1990s. In the meantime, 10 years of missed chances for local agricultural growth occurred.

Lesson 4: The subsystem of associations and cooperatives must develop autonomy. The public services have substantial technological capabilities, institutional, and legal legitimacy, and the authority to oversee, manage, and assess. Cooperatives and organizations possess significant technical capabilities and are situated in close proximity to farmers and communities. However, they are subject to public regulation and are financially highly dependent, which limits their autonomy. Establishing horizontal alliances and fostering synergistic action within them has the potential to stimulate creativity and contribute to the creation of sound development programs.

SUMMARY

According to an analysis of the IPM-FFS experiences in Indonesia and the Philippines, the FFS approach to large-scale farmer knowledge delivery is vulnerable to the same risks of fiscal unsustainability as other large-scale extension initiatives where real results have been unsatisfactory thus far. Similar to the Philippines, an FFS program might serve as the cornerstone of a country's agricultural extension system, and in such circumstances, its viability would depend on public funding. However, only a small number of farmers can be trained using the FFS strategy in the Philippines due to the high cost per farmer and the small amount of funding available for extension. Some may suggest using farmer-led field schools as a solution to this financial conundrum because they transfer a portion of the costs to the farming community. But Indonesian experience indicates that a substantial training effort under the FFS paradigm cannot be sustained just through farmer-led field-school activities. The study's conclusions are highly pertinent to the debate over comparable extension initiatives in other areas. Our two-country study indicates that if the goal is to scale up these activities based on the pilots, then there is a need for extreme selectivity and caution when starting FFS operations, with an emphasis on the economic sustainability of programs. The same prudence should be used with any extension initiative that is anticipated to receive government funding and has significant ongoing costs. When assessing the benefits of proposed projects, it would be helpful to carefully examine predicted public expenditures over time in relation to the likely budgets of the relevant public agencies and local governments.

BIBLIOGRAPHY

1. AED (Academy for Educational Development). (1985). *Communication strategies for agriculture: Hybrids of a different kind*. Washington, DC. AED.
2. Alex, G., & Rivera, W. (2004). *Demand driven approaches to agriculture extension Case Studies of International Initiatives* (Vol. 3).
3. Amanor, K., & Farrington, J. (1991). *NGOs and Agricultural Technology Development*. In (W. M. Rivera and D. J. Gustafson, eds.). *Agricultural Extension: Worldwide Institutional Innovation and Forces for Amsterdam: Elsevier.*
4. Ameur, C. (1994). *Agricultural extension. A step beyond the next step* (Vol. 23). World Bank Publications.
5. Anderson, J., & Van Crowder, L. (2000). The present and future of public sector extension in Africa Contracting out or contracting in *Public Administration and Development the International Journal of Management Research and Practice*, *20*(5), 373–384.
6. Antholt, C. H. (1991). *Agricultural extension in the 21st century Lessons from South Asia. Worldwide Institutional Innovation and Forces for Change (W. M. Rivera and D. J. Gustafson, eds.). Amsterdam: Elsevier.*
7. Antholt, C. H. (1994). *Getting ready for the twenty-first century Technical Change and Institutional Modernization in Agriculture. World Bank Technical Paper 217. Washington D.C.: World Bank.*
8. Ashby, J. (2003). Local agricultural research committees (CIALS). In *Investment note*. Washington, DC. The World Bank.
9. Ashby, J. A., Braun, A. R., Gracia, T., Guerrero, M. D. P., Hernández Romero, L. A., Quirós Torres, C. A., & Roa Velasco, J. I. (2000). *Investing*

in Farmers as Researchers Experience with Local Agricultural Research Committees in Latin America.

10. Axinn, G. H. (1988). *Guide on alternative extension approaches: FAO, Rome.*
11. Bebbington, A., & Kopp, A. (1998). Networking and rural development through sustainable forest management. Frameworks for pluralistic approaches. *UNASYLVA FAO*, 11–18.
12. Benor, D., & Harrison, J. Q. (1977). *Agricultural extension. The training and visit system. Washington, D.C.: World Bank.*
13. Berdegué, J. (2000). *Systems-oriented agricultural extension and advisory services for small farmers in Latin America. Santiago, Chile. RIMISP.*
14. Berthe, A. (2015). Extension and advisory services. Rural extension services for agricultural transformation. *Background technical paper Feeding Africa*, 21–23 October 2015.
15. Bingen, R. J. (1998). Cotton, democracy, and development in Mali. *The Journal of Modern African Studies*, *36*(2), 265–285.
16. Birkhaeuser, D., Evenson, R. E., & Feder, G. (1991). The economic impact of agricultural extension. A review. *Economic Development and Cultural Change*, *39*(3), 607–650.
17. Braun, A. R. (1997). An analysis of quality in the Indonesian integrated pest management training project. *Report of a technical audit conducted for the World Bank of the Indonesia Integrated Pest Management Training Project*. Washington, DC The World Bank.
18. Byrnes, K. J. (2001). Farmer organizations. Tapping their potential as catalysts for change in small-farmer agricultural systems. In *Agricultural extension systems. An international perspective*. North Chelmsford, MA Erudition Books.
19. Campbell, D., & Barker, S. C. (1997). Selecting appropriate content and methods in programmer delivery. A Reference Manual. FAO: Rome.
20. Carney, D. (1998). *Changing public and private roles in agricultural service provision., viii + 90 pp.*
21. Chamala, S., & Shingi, P. M. (1997). Establishing and strengthening farmer organizations. Chapter 21, In: Swanson, Burton, E., Bentz, Robert P., & Sofranko, Andrew J. (ed) Improving Agricultural Extension: A Reference Manual, FAO, UN, Rome.
22. Collinson, M. (Ed.). (2000). *A history of farming systems research*. CABI.
23. Cristóvão, A., & Portela, J. (1995). *Desenvolvimento rural em Trás-os-Montes: Contributo para a análise crítica das intervenções dos últimos 20*

anos. Comunicação ao III Encontro Nacional da Associação Portuguesa de Desenvolvimento Regional. Porto. APDR.

24. Crowder, L. V. (2000). National agricultural advisory service (NAADS). *FAO/World Bank Formulation Mission Report*, Uganda.
25. Currle, J., & Schütz, P. (2000). Privatizing agricultural extension services in two new German federal states. Necessary conditions emerging from experience. In (pp. 130–140). Rome Food and Agriculture Organization (FAO).
26. Davis, K. E. (2009). Agriculture and climate change: An agenda for negotiation in Copenhagen for food, agriculture, and the environment. The important role of extension systems. *IFPRI Focus 16 Brief II.*
27. Dillinger, W. R. (1995). *Decentralization, politics and public services* (No. 15635, pp. 1–0). The World Bank.
28. Eicher, C. K. (2001). Africa's unfinished business. Building sustainable agricultural research systems.
29. FAO (Food and Agriculture Organization of the United Nations). (1997). *Improving agricultural extension: A reference manual*. Rome: FAO.
30. FAO (Food and Agriculture Organization of the United Nations). (1990). *Global consultation on agricultural extension*. Rome FAO.
31. FAO. (1987). *Marketing extension services for small farmers*. AGSM Occasional Paper No. 1. Rome Agricultural Services Division, Marketing and Rural Finance Service.
32. FAO. (2000). *Farm Net* [Leaflet]. Rome Sustainable Development Department, Communications for Development.
33. FAO/World Bank. (2000). *Knowledge and information systems for rural development (AKIS/RD) Strategic vision and guiding principles*. Rome FAO/World Bank.
34. Farrington, J. (1997). *The role of nongovernmental organizations in extension. In: Improving Agricultural Extension: a reference manual (Eds: B.E. Swanson, R.P. Bentz, & A.J. Sofranko). Rome: FAO.*
35. Feder, G., & Slade, R. (1993). Institutional reform in India. The case of agricultural extension. *In: The Economics of Rural Organizations (K. Hoff, A. Braverman, and J. Stiglitz, eds.). Oxford: Oxford University Press.*
36. Feder, G., Birner, R., & Anderson, J. R. (2011). The private sector's role in agricultural extension systems. Potential and limitations. *Journal of Agribusiness in Developing and Emerging Economies*, *1*(1), 31–54.

37. Feder, G., Willett, A., & Zijp, W. (2001). Agricultural extension. Generic challenges and the ingredients for solutions. In *Investing in rural extension Strategies and goals* (pp. 313–353). Springer US.
38. Fei, X., & Hiroyuki, T. (2000). Fee-charging extension in China: Implications and function. *Journal of Extension Systems*, *16*(1), 58–67.
39. Food and Agriculture Organization of the United Nations. (1999). *Technical assistance to the integrated pest management training project. Indonesia project findings and recommendations. Rome. FAO.*
40. Fresco, L. O. (2000). Scientific and ethical challenges in agriculture to meet human needs. *Food Nutrition and Agriculture*, (27), 4–13.
41. Garfield, E., Guadagni, M., & Moreau, D. (1996). Colombia: Decentralization of agricultural extension services. In *Extension workshop alternative mechanisms for funding and delivering extension*. The World Bank Washington, DC.
42. Griffis, D. (1996). ADAS (United Kingdom). In *Extension workshop alternative mechanisms for funding and delivering extension*. Washington, DC World Bank.
43. Grosskopf, W. (1989). *Das bisherige staatliche Konzept reicht nicht aus. WWL/BLW*, 11, 89.
44. Gruber, I. (2002). *Rinderspezialberatung in Baden-Württemberg* [Doctoral dissertation]. Thesis, elaborated at the University of *Munich Weihenstephan*, unpublished.
45. Gustafson, D. J. (1991). *The challenge of connecting priorities to performance. One state's response to the forces for change in US extension. In: Agricultural Extension: Worldwide Institutional Innovation and Forces for Change (W. M. Rivera and D. J. Gustafson, eds.). Amsterdam: Elsevier.*
46. Hanson, J. C., & Just, R. E. (2001). The potential for transition to paid extension. Some guiding economic principles. *American Journal of Agricultural Economics*, *83*(3), 777–784.
47. Hercus, J. M. (1991). *The commercialization of government agricultural extension services in New Zealand.*
48. Hoffmann, V. (2002). *Beratung landwirtschaftlicher Betriebe: Bund und Länder weiterhin in der Pflicht. German Journal of Agricultural Economics*, *51*(7), 329–331.
49. Hoffmann, V., Lamers, J., & Kidd, A. D. (2000). Reforming the organization of agricultural extension in Germany Lessons for other countries.
50. Howell, J. (1986). Accountability in extension work. in G.E. Jones (ed.)

Investing in rural extension: Strategies and goal. London: Elsevier.

51. Howell, J. J. (1985). *Recurrent costs and agricultural development. Overseas Development Institute, London, p. 223.*
52. Humphries, S., Gonzales, J., Jimenez, J., & Sierra, F. (2000). Searching for sustainable land use practices in Honduras: Lessons from a programme of participatory research with hillside farmers. *Network Paper/Overseas Development Institute, Agricultural Research and Extension Network*, 104.
53. Israel, G. D. (1982). *The training and visit system. A warning* (Projects Advisory Staff Paper). Washington, DC. World Bank.
54. Kaimowitz, D, W. M. Rivera & D. J. Gustafson (1991). The evolution of links between research and extension in developing countries. In *Agricultural extension. Worldwide institutional innovation and forces for change*. Amsterdam Elsevier.
55. Kidd, A. D., Lamers, J. P. A., Ficarelli, P. P., & Hoffmann, V. (2000). Privatizing agricultural extension: Caveat emptor. *Journal of Rural Studies*, *16*(1), 95–102.
56. Koch, L. (2001). Zukunft der landwirtschaftlichen Beratung.
57. Lionberger, H. F. (1968). *Adoption of New Ideas and Practices*. Ames Iowa State University Press.
58. Maguire, C. J. (2000). From agricultural education to education for rural development and food security. In *Fifth European conference on higher agricultural education*, Plymouth, UK. Retrieved from www.fao.org/sd/EXdirect/EXre0029.htm. (accessed on 17th October 2024
59. Menter, H. (2001). Changing institutions: A study of the process of institutionalization of CIAL methodology in Corpoica. *London School of Economics*. Paper prepared for IPRA, CIAT, CALI, Columbia.
60. Nagel, U. W. (1997). Alternative approaches to organizing extension. A reference manual, edited by Swanson BE, Bentz RP, Sofranko AJ. Food and Agriculture Organization of the United Nations; Rome.
61. Ngala, A. M. (2000). Farm management in extension services. A brief on Zambia. Rome.
62. Oakley, P. (1986). The monitoring and evaluation of non-material objectives of extension. Pages 247-253 in Gwyn E. Jones, ed. Investing in rural extension: Strategies and goals. Elsevier Applied Science Publishers, New York.
63. Obibuaku, L. O. (1983). *Agricultural extension as a strategy for agricultural transformation. University of Nigeria Press p 81.*

64. OECD. (1999). *Summary and evaluation of main developments and changes in organizational forms of and approaches by the AKS in OECD member countries*. Paris OECD. Directorate for Food. Agriculture and Fisheries. Committee for Agriculture.

65. Portela, J., & Cristóvão, A. (1991). *Proagri, extensão e desenvolvimento rural: Contributo para uma reflexão. Economia e Sociologia,* (52), 43–74.

66. Porter, G. (2001). Accountability, subsidiarity, diversity. http://www. converge. org.nz/pirm/subsidia.htm

67. Pretty, J. N., & Vodouhe, S. D. (1997). Using rapid or participatory rural appraisal. In Improving Agricultural Extension, edited by Swanson B.E, Bentz, R.P. & Sofranko, A.J. Rome, FAO: United Nation.

68. Purcell, D., & Anderson, J. R. (1997). *Agricultural extension and research. Achievements and problems in national systems*. World Bank Publications.

69. Qamar, M. K. (2000). Agricultural extension at the turn of the millennium. Trends and challenges.

70. Quizon, J., Feder, G., & Murgai, R. (2000). A note on the sustainability of the farmer field school approach to agricultural extension. *Development Economics Group*. The World Bank. Washington, DC.

71. Rivera, W. M. (1991). *Sustainable agriculture. A unifying concept for agriculture and a central strategy for extension.*

72. Rivera, W. M. (2000). The changing nAQ: Kindly provide additional information for this reference.

73. ature of agricultural information and the conflictive global developments shaping extension. *The Journal of Agricultural Education and Extension*, *7*(1), 31–42.

74. Rivera, W. M., & Cary, J. (1997). *Privatizing agricultural extension ; In: Improving Agricultural Extension: a reference manual (Eds: B.E. Swanson, R.P. Bentz, & A.J. Sofranko). Rome: FAO.*

75. Rivera, W. M., Zijp, W., Alex, G., Ashworth, V., Crowder, L. V., & Anderson, J. (2000). *Contracting for extension: Review of emerging practices*.

76. Röling, N. (1986). *Extension and the development of human resources. The other tradition in extension education. In Gwyn E. Jones' (ed.), Investing in Rural Extension: Strategies and Goals London and New York: Elsevier.*

77. Rouse, J. (1999). Financing economic self-reliance and member participation in farmer organizations in Kenya. In *ILO, International Training Centre of the ILO, Grassroots' Organizations, Decentralization and Rural Development.*

78. Russell, J. F. (1986). Extension strategies involving local groups and their participation, and the role of this approach in facilitating local development.

79. Sánchez-Griñan, M. I. (1998). *Seguridad alimentaria y estrategias sociales: Su contribución a la seguridad nutricional en areas urbanas de América Latina. Washington, DC: Inst. Internac. de Investigaciones sobre Politicas Alimentarias.*

80. Shah, A. (1994). *The reform of intergovernmental fiscal relations in developing and emerging market economies*. The World Bank.

81. Swanson, B. E. (1997). Changing paradigms in technology assessment and transfer. *Unpubl. Urbana, IL. University of Illinois. INTERPAKS.*

82. Swanson, B. E. (Ed.). (1984). *Agricultural extension. A reference manual* (2nd ed.). Rome. FAO.

83. Tacken, W. (1996, June). The Dutch extension service. The change from a publicly funded extension service to a privatized, client-oriented organization. In *Extension Workshop Alternative Mechanisms for Funding and Delivering Extension*.

84. Tanaka, H. (2001). *The New Paradigm for the Community Forestry Research and the Implication to the Extension System: Lesson learning from Farmer Forest Management School*. Rome FAO, Community Forestry Unit, Policy and Institution Branch.

85. Tobar, L. G. (1996). CREA AACREA associative private agricultural extension in Argentina. In *Extension Workshop Alternative Mechanisms for Funding and Delivering Extension*. Washington, DC. World Bank.

86. Umali-Deininger, D., & Schwartz, L. A. (1994). *Public and Private Agricultural Extension: Beyond Traditional Frontiers* (Vol. 236). World Bank Publications.

87. Umali, D. L. (1996). New approaches to an old problem The public and private sector in extension. In *Extension workshop. Alternative Mechanisms for Funding and Delivering Extension* (SA2). The World Bank. Washington, DC.

88. Van Crowder, L. (1996). Decentralized extension. Effects and opportunities. http://www.fao.org/sd/Exdirect/EXan0013.htm

89. Van de Fliert, E., Pontius, J., & Röling, N. (1995). Searching for strategies to replicate a successful extension approach: Training of IPM trainers in Indonesia. *European Journal of Agricultural Education and Extension*, *1*(4), 41–63.

90. Van den Ban, A. W. (1986). Extension policies, policy types, policy formulation and goals. In *Investing in rural extension: Strategies and goals* (pp. 91–97). Elsevier.
91. West, E. G. (1996). *Education vouchers in practice and principle A world survey*. Washington, DC: World Bank.
92. Wete, F. N. (1991). New technology for transferring agricultural information. In: Rivers WM, Gustafson DJ (Eds.) Agricultural Extension: World-wide Institutional Evolution and Forces for Change. Elsevier, New York.
93. Wiebers, U. C. (1993). *Integrated pest management and pesticide regulation in developing Asia. Papers 211, World Bank - Technical Papers.*
94. Williams, J. R., Jones, C. A., & Dyke, P. T. (1984). A modeling approach to determining the relationship between erosion and soil productivity. *Transactions of the ASAE*, *27*(1), 129–144.
95. William M. Rivera, M. Kalim Qamar and L. Van Crowder (2001). Agricultural and Rural Extension Worldwide: Options for Institutional Reform in the Developing Countries.
96. Willige, B. (1999). Kraich-Cow report. Sinsheim.
97. Wilson, M. (1991). Reducing the costs of public extension services. Initiatives in Latin America.
98. World Bank. (1990). *Agricultural extension. The next step. Policy and Research Series No. 13*. Washington, DC.
99. World Bank. (1997). *Rural development. From vision to action. Sector Strategy. Environmentally and Socially Sustainable Development Studies and Monographs Series 12*. Washington. DC.
100. Zijp, W. (1994). *Improving the transfer and use of agricultural information. A Guide to Information Technology* (Vol. 247). World Bank Publications.

INDEX

H

I

J

K

L

M

N

O

P

Q

R

S

T

V

W

Y

FSC
www.fsc.org
MIX
Papier | Fördert
gute Waldnutzung
FSC® C083411

Kolibri 360 GmbH
Ferdinand-Jühlke-Straße 7
99095 Erfurt, Deutschland
produktsicherheit@kolibri360.de